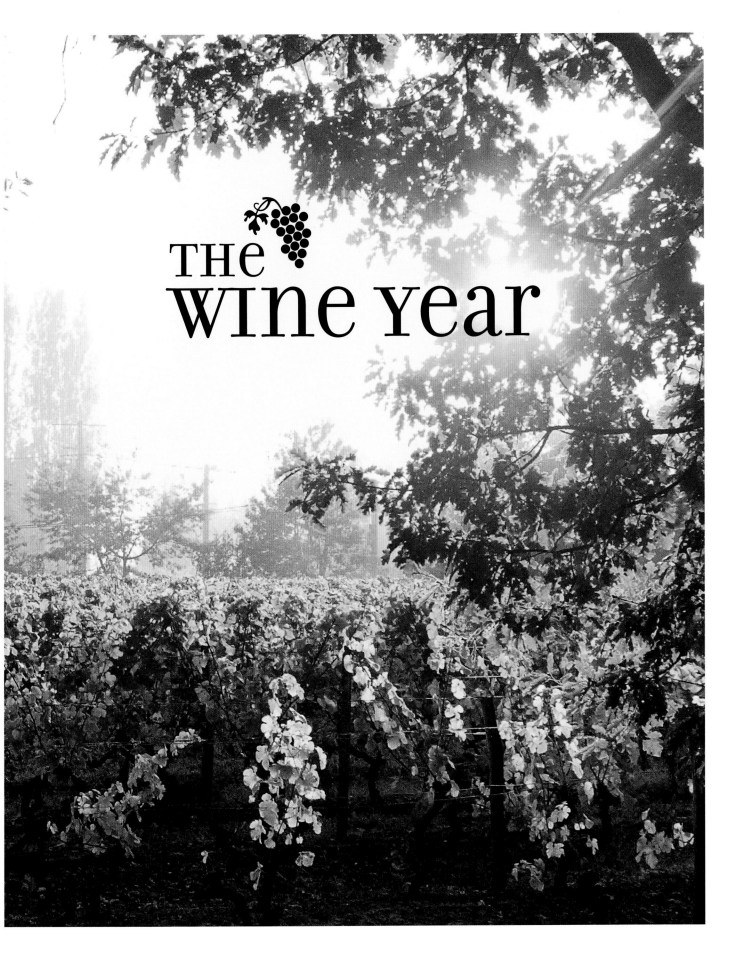

Rosalind Cooper

THE wine year

MERRELL

LONDON · NEW YORK

Introduction 7

Introduction

··

High-trained vines lie
dormant in winter in the
Trentino Alps of Italy.

Wine is a product of the seasons. From the earliest signs of growth in spring, through the crucial flowering period of early summer, to the final fruition at harvest in the autumn, the vine's fate is tied to this annual cycle. This book celebrates the monthly round in the vineyard and augments this with a feast of fact and entertainment related to food as well as wine. There is no doubt that learning more about what is in the bottle adds another dimension to tasting and drinking wine; and pairing wine with the right dishes makes food, drink and life more enjoyable.

The structure of this book closely follows that of my previous book on wine, *The Wine Almanac* (1986). Over the intervening twenty-plus years, there has been a fundamental shift in the wine trade. Where once France, Germany and Italy held sway over every wine list, today you are likely also to find wines from California, Australia or Chile in any bar or pub. The secret to the success of these newcomers is precisely that they are designed to appeal to the widest possible audience. With their appetizing fruitiness, reliable quality and absence of any tannic harshness on the finish, they are now the key wines of the modern world. The overwhelming popularity of Australian Shiraz has resulted in producers in the South of France renaming some of their red table wines – from Syrah to Shiraz in one generation. There are similar stories across all of the traditional European wine regions, as producers rush to catch up with the latest trends.

The wide, diverse world of wine is also open for business, and wine tourism – visits to vineyards, wineries and restaurants, and organized tours to wine fairs and festivals – is a fascinating and satisfying way to access the full story behind the bottle. There is nothing to equal sitting at a table with an estate proprietor, listening to a passionate account of a long family history and devotion to making fine wines from the most appropriate, and healthiest, grape varieties. Likewise, tasting wines with a knowledgeable guide, and visiting the cellars before savouring a picnic amid the vines, is an unforgettable experience. Whether your journey is long-planned, completely spontaneous (a sudden diversion from a dull business trip, perhaps?) or just a dream at this stage, this book will provide inspiration and guidance.

How to Use This Book

The book is divided into twelve chapters that trace the months of the year, both in the vineyard and in your own home; offer suggestions about places to go to discover more about wine at first hand; and advise on how to combine specific foods and wines to delicious effect. Each chapter includes the following sections:

Knowledge

To start the chapter, there is a feature that relates to the month under consideration, and helps to improve your background knowledge of wine in general. From how grapes are nurtured to making the ultimate wine selection for your personal cellar, there is a wealth of wisdom and advice. This section also provides the answers to such significant questions as 'Is wine good for you?' and 'What is organic wine?'

Entertaining

Sharing wine with friends and family is one of life's richest pleasures, and this enjoyment connects us to all our wine-loving ancestors. Each chapter offers suggestions for how to use wines, from celebrating the New Year to getting the most from a well-chosen home cellar. Along the way there are plenty of ideas and recipes for meals, from the grand dinner to the most modest barbecue, all matched with just the right wines from around the world.

Travel

This section aims to inspire you to travel to vineyards to learn about the varied places in which wine is made and where to relish their atmosphere. From the cool climate, slopes and limestone cellars of Champagne, via the fierce heat of central Spain or South America, to classical Tuscany or the autumn colours of upper New York State, the travel features grant revealing insights into the kaleidoscope of the wine world, and give you some understanding of what awaits you in the enticing vineyards of both the Old and the New Worlds. Some of the exciting and amusing wine-related fairs and festivals that take place each month are also featured.

Interview

Each chapter concludes with one or more exclusive question-and-answer sessions with key players from the modern world of wine, giving an insight into the wine business and some of its complexities, as well as regional differences in taste and style.

Finally, do take a look at the useful glossary of wine terms at the back of the book, and give yourself some jargon to use next time you are tasting or discussing a bottle of wine. There is also a useful list of key websites to consult, covering many aspects of wine and the main regions of production. Using this as a source, you should be able to track down most of the wines mentioned in this book.

This is an exciting time for wine and its producers. As the consumer becomes more demanding and knowledgeable, the quality of wine has never been better; and it is being made in all sorts of locations, from rural China to chilly Patagonia. There is a big world of wine out there, and your journey starts with this book.

Scenes from South Africa (top left and right), Chile (centre) and France (bottom left) exemplify the international character, history and agricultural nature of wine.

Ripe Australian red grapes
stain oak barrels in the
Penfolds cellars.

January

Growing Great Grapes

In the northern hemisphere, January falls in the depths of winter; in the southern hemisphere, the vineyard's winter work begins in July. At this point in the winter the vine is dormant, and this is the moment for pruning. It is an arduous process, often still carried out by hand, and demands great skill, patience and endurance in the cold weather. Why is pruning the key to growing great grapes?

Pruning is certainly not a modern practice. There are references in the Bible to avoiding pruning on the Sabbath, and the famous 'they shall beat their swords into ploughshares' is followed by 'and their spears into pruning hooks' (Isaiah 2:4). There are various aspects to pruning, including making a practical shape for the vine so that it can be tended during the year; but the principal intention is to curb the vine's natural enthusiasm for creating bunches of grapes, and restrict the number that it produces. The result is a higher quality and lesser quantity of grapes, thereby enabling the wine to be improved.

In winter the vines have lost their leaves and are reduced to bare branches, known as canes, ready to be trimmed back. If the process is left too late, it is quite possible to damage the vine as the sap rises, and to create serious problems for the coming season. It is also

ABOVE AND RIGHT: In the depths of winter, snow and ice form a protective layer over the dormant vine, prior to pruning. In spring, frost becomes an enemy as it can curtail early growth in vineyards such as these in Bordeaux.

OPPOSITE: In a Champagne cellar, special racks called *pupitres* are designed to facilitate the removal of sediment from the wine.

important to time pruning so that severe frost cannot wreak havoc on the vine, which is why one often sees pruners at work in winter fog and damp – less harmful to the plant than clear, bright, frosty days. The way vines are pruned owes a good deal to tradition and now also to various legal requirements in those regions of the world that regulate their wines: for instance, the French Appellation Contrôlée (AC), which dictates precise yields per hectare of vines for wines seeking their AC. Perhaps the simplest method of vine training is the freestanding bush shape, known in France as the *gobelet* and common in many older established wine regions. Vines trained in this way withstand wind and scorching sun well; visit the Priorat area of Catalonia in Spain, for instance, to see some magnificent old vines in this form, a source of some exceptional, deep-red wines.

More familiar are rows of vines trained along post and wire at different heights and in slightly differing styles, depending on the region. It is practical to manage vines in this way: they can be trimmed easily, often using machines; the land between the rows can be ploughed or fertilized; and then, eventually, grapes may be picked using a mechanical harvester. These days, the use of intensive chemical treatment is increasingly frowned upon, and many producers are doing their utmost to avoid the use of herbicide (see 'April', pp. 59–62, for more about organic and biodynamic viticulture), instead planting grass or what is known as 'green manure' between the rows of vines. Green manure includes such plants as mustard or clover, which may be ploughed in to improve the quality of the vineyard soil without encouraging excess leaf growth.

Many vineyards plant grass between the rows. This is very helpful as a moisture retainer, or to avoid erosion on a steep slope, but it does require maintenance, by mowing with a small tractor or, in some regions, allowing sheep to graze between the rows of vines. The latest concept is to breed miniature sheep that will be unable

to reach up and nibble the grapes once they ripen; these are known as 'Babydoll' versions of the well-known Southdowns, originally from the UK. They would certainly add to a winery's tourism appeal, and wine tourism is a vital part of income for many producers the world over.

What are the other keys to getting the best possible grapes? Constant surveillance is essential, with regular visits to the vines. As spring progresses, the prospect of a sudden frost is more threatening, and the worst fear of all is a sharp frost or hail in May and June (in the northern hemisphere), or, in the southern hemisphere, in November and December. At these times, when the vine is in flower, the nascent bunches may be literally knocked off the vines, never to reappear, and the loss of the crop may be total. There are many ways to deal with frost, from what Americans endearingly call 'smudge pots' – braziers that create warm patches amid the vines and prevent the air chilling right down – to spray systems that coat each vine with water and 'seal' them up, and even

heated electric coils running along the rows, effectively acting as 'electric blankets' for very valuable vineyards.

Then there is a catalogue of pests and other threats to the vine. Animals, including deer, rabbits and mice, and all types of bird, love to feast on the grapes when these are ripe, and nibble the leaves. In many regions, fencing is essential, and this is costly. Harder to spot are such enemies of the vine as red spider mite, beetles and various moths. There are some cunning organic methods used to combat these pests, such as the method known as *confusion sexuelle*, whereby pheromones that confuse and prevent moths from breeding are unleashed from small phials attached to the vines. Chemical treatment cannot, however, always be avoided.

Smaller even than mites or moths, but just as dangerous, are the moulds that are devastating to the vine. Spraying with so-called Bordeaux mixture (copper sulphate and hydrated lime in solution) is the most common remedy, generally considered to be fairly non-invasive as far as the grape is concerned. This gives the characteristic grey-blue colour seen on so many vine posts and buildings in wine regions. There is one mould, however, that is more than welcome in certain regions where sweet dessert wines are produced. Sauternes is perhaps the most famous example of a wine affected by *Botrytis cinerea*, the 'noble rot', which looks appalling – a grey fur all over the bunches – but lends the wine a uniquely smoky, musky appeal that is enchanting. The mould pierces the skin of the grapes and the resulting loss of liquid by evaporation means that the juice yields higher sugar levels. It is not uncommon for pickers to pass through the rows of vines at such fine vineyards as Château (Ch.) d'Yquem at least three times, each time selecting only the most 'rotten' bunches and even single grapes. ('Château', often abbreviated to 'Ch.', is the term used for wine estates, whether these are extensive or simply smallholdings with their own cellars.)

Finally, there is the great, unsolved irritation of phylloxera. This pest, a type of sap-sucking insect that nibbles the roots and eventually kills the vine, has plagued vineyards in almost all parts of the wine world since the late nineteenth century. There are still prizes on offer dating from the French government of that era for anyone who can conquer this menace with science. To date, the only remedy is to graft the European grape varieties (*Vitis vinifera*) on to sturdier native American rootstock, which is impervious to phylloxera.

In all cases, quick reactions are essential to the winegrower. There is no doubt that 'the price of ripening is eternal vigilance', to paraphrase the famous quotation, often attributed to Thomas Jefferson.

LEFT: Challenging growing conditions in high-altitude Italian vineyards can add cost to wine production.

BELOW: The *Botrytis cinerea* mould ('noble rot') on these Bordeaux grapes results in increased sugar levels.

Entertaining with a Sparkle

In most countries, the new calendar year begins in January, and in many parts of Europe it is welcomed with a real feast, which can see festivities lasting throughout the night and into the morning.

It might seem wildly extravagant to create an entire meal based on one style of wine, but take a completely new look at Champagne (strictly, sparkling wine produced in the Champagne region of France) or other sparkling wines with food, and think like the French. They actually plan menus to complement this style of wine, and manage to sit through five or six courses only to arise relaxed and mellow. Contrary to popular myth, sparkling wines can complement all types of cuisine to great effect, and the relatively low alcohol levels in most of these wines is a boon when compared to some table wines. What could be a more pleasing sight than chilled ice buckets replete with bottles of fizz, and a glorious array of dishes to come? Or a table with handwritten menus that demonstrate how a sparkling wine may be found to match any dish? If money is no object, take a tip from the Champagne families themselves, and serve their wonderful wines throughout the meal; or make an informal occasion special with the addition of fizz.

Always pour Champagne slowly and steadily, to avoid excess frothing in the glass. This glass displays the classic *fine mousse*, the small, persistent bubbles of a quality sparkling wine.

French Grand Style

∿

If you would like to conjure up visions of elegance – butlers wearing white gloves serving iced Champagne – then set your table in as classic a style as you can manage, with immaculate white or palest pink linen, gleaming silverware, white crockery and fine crystal flutes for the sparkling wine. Add fresh flowers, set in low silver bowls so that guests are able to talk freely over them. Simple white candles set in silver or crystal holders would be ideal for an evening event, as they reflect the glitter of the *fine mousse* of bubbles in the glass.

The menu should be equally classic and not too fussy, to allow the wines to show to advantage. How about starting with smoked fish: salmon, eel or trout, with just the very lightest of horseradish dressing mixed with yogurt, crème fraiche and chopped chives? The crispness of a light Brut (dry) or even Extra Brut (bone dry) non-vintage Champagne would cut through the oily fish and create excellent balance. Serve with sliced brioche; its slight sweetness highlights the toasty flavour found in all good Champagne.

Next, a very simple roast leg of lamb, the lean cut called *gigot* in France, served with a light redcurrant sauce made with seasonal fruit. New potatoes with butter and some grilled courgettes (zucchini), cut into strips and basted with a little oil, add nuttiness, which would match perfectly with a vintage pink Champagne.

Choose fresh goat's cheese for the next course, as this is a great match for dry non-vintage Champagne, especially Blanc de Blancs (made with 100 per cent Chardonnay). Or be daring and offer more pink Champagne with a really ripe Brie; they come from neighbouring regions of France and have an excellent affinity.

For a pretty pudding, make jelly with pink Champagne, using a decorative mould. Dip bunches of redcurrants, raspberries and strawberries into sugar syrup to glaze, and scatter these round the turned-out jelly. Serve with light cream to bring out the distinct creamy quality of an off-dry Champagne; look for Extra Dry styles, which are not in fact very dry, and have just enough depth of flavour to make excellent dessert wines.

A Sparkling Supper

On a more informal note, and when budgets are not limitless, try this combination:

As an aperitif, serve some fine sparkling wine, such as a Sauvignon Blanc from New Zealand; this type of wine has a distinctive fruity style and is a great match with little nibbles, such as tiny fresh cherry tomatoes stuffed with creamy cheese, or slivers of good-quality ham wrapped round asparagus tips.

Take advantage of the way sparkling wine complements chilli and fairly hot or spicy food by preparing either a Spanish paella with fish and chicken in saffron rice, or some couscous with spicy vegetables. Serve with salad, but avoid too much vinegar; instead use lemon juice or *verjus* (verjuice) – fresh, tart grape juice – if you can find it, with some good, light olive oil. Serve with a dry Cava, or a sparkling wine from California, such as Domaine Chandon or Schramsberg.

Next, why not try matching some cheese and fruit to a rich, ruby-tinted Australian sparkling Shiraz? This unusual, zingy wine has great depth of flavour and is delectable with hard cheese, such as Cheddar, Cantal or Gouda, served with oatcakes, wheat thins and a fruity chutney. Alongside, offer a platter of sliced seasonal fruits, such as plums and pears. For those with a sweeter tooth, open a bottle of off-dry sparkling Sekt from Germany, or perhaps a French Blanquette de Limoux. For an educational and delicious twist to the meal, offer some sugar almonds and candied fruit to see how their flavours may also be found in these fine sparkling wines.

Visiting the Vineyards of Champagne

FRANCE

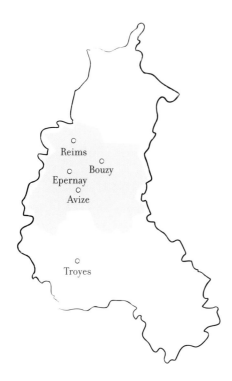

Reims
○
○
Bouzy
○
Epernay
○
Avize

○
Troyes

The map above shows the French *département* (administrative district) of Champagne; the coloured area within it shows the region's main Champagne-production area.

It would be difficult for the landscape of the Champagne region of France to live up to the glamour that surrounds its marvellous, vinous namesake. A bottle of Champagne epitomizes elegance, charm and style, whereas the countryside of its birth, although pretty enough in a rustic way, is not spectacular. The hills are gently rolling, there are dense forests, the plains extend along the valley of the Marne and, because this is northern France, the prevailing light is frequently filtered and grey rather than brilliant. So the atmosphere can seem a little austere, almost melancholy, rather than celebratory. In the two major towns of the region, Reims and Epernay, the headquarters of the famous Champagne houses stand four-square behind tall railings, and the main impression is of solid bourgeois comfort rather than opulence. Despite the fact that many of the companies warmly welcome visitors, this is not always apparent to the casual tourist.

So the best idea is to do some research before setting out, or choose to join one of the many organized tours that visit the region. Asking an established wine merchant for a personal recommendation or introduction to one of the Champagne houses is also a sensible way to ensure access on arrival. A visit to one of the vast, cobwebbed limestone *caves* (cellars) that lie beneath the streets of Reims and Epernay is unforgettable. It also gives some notion of how much effort goes into creating this most complicated wine. Many cellars stretch for literally miles and store millions of bottles in these ideal, humid conditions.

From an initial fermentation in conventional vats or – for a few diehards, such as Krug and Bollinger – wooden barrels, the wines are blended with great skill to achieve a consistent 'house style', essential in non-vintage Champagne. Only a very few wines, from good years, are sold with a vintage date. All wines, vintage or not, go through a secondary fermentation in individual bottles; yeasts and natural sugars are added to the wine, and the bottle sealed with a crown cap as a beer bottle would. Once this process is completed, the resulting fizz (the CO_2 is trapped in the bottle after fermentation) rests on its lees (the deposit left in the bottle) for a remarkably long time – in the case of a vintage wine, perhaps three years. Then the bottles are gently twisted and shaken daily in a system known as *remuage* (invented by none other than the Veuve, or widow, Clicquot at her kitchen table). Once, this was a highly skilled hand craft, but today most houses use sophisticated machinery in a

process that eventually leaves a 'plug' of sediment in the neck of the bottle. This plug is expelled immediately prior to the Champagne being corked. Curiously, the corks are straight-sided at first, and take on the familiar mushroom shape over time as a result of being compressed to fit into the bottle. It is interesting to note that there is as much pressure in a bottle of Champagne as in the tyre of a London bus. The tiny amount of fizz lost when the bottles are opened is replaced by what is known as the *dosage* – a careful mix of old wine and grape sugars to create Brut (dry), Extra Dry (off-dry) or Doux (fairly sweet) styles according to taste.

All of this and more is demonstrated when you take a tour, and then of course comes the added treat: the tasting. Some houses will charge for this, as Champagne is very expensive to make, store and market; even a small glass is of high value.

These chalky cliffs in the Ay area of the Champagne region are the source of excellent Chardonnay. Limestone is at the heart of many other fine wines, such as Chablis and Burgundy.

The region is easily reached by car from other parts of Europe, and from England by Eurostar train. Wily Champagne executives from Krug or other top houses leave the train at Lille and continue on from there by car, meaning their overall journey is absurdly swift. Taking a car all the way does of course have its own advantages: finances permitting, you can seize the chance to stock up on some cases of carefully selected Champagne for your home cellar.

For not all Champagne comes from the famed Champagne houses based in Reims and Epernay. Even in those two centres, there are smaller and perhaps better-value establishments that sell much of their produce to French customers, and that therefore present a welcome discovery for the foreign visitor. More exciting still is a visit to a small-scale family producer, many of whom are based in the vast Marne Valley. Here they still grow the classic three grapes of Champagne – Pinot Noir, Pinot Meunier and Chardonnay – but make distinctive and characterful fizz that will probably be unlike anything you have tried before.

Although Champagne is generally a white wine, it is often made with a high percentage of juice from red grapes,

LEFT: Krug is a connoisseur's Champagne, made in small quantities for the discerning drinker.

RIGHT, TOP: The skilled hands of a *remueur* gently turn Champagne bottles. Specialized machines have now largely replaced this human element.

RIGHT: Pinot Noir grapes are worked into an old-fashioned *pressoir* that will extract their white juice to produce the wine, which will become Champagne after a secondary fermentation.

La plus ancienne Maison de Vins de la Champagne. Aÿ 1584

ABOVE: A striking mural at the cellars of Champagne Gosset in the village of Ay, showing how the cellars extend under the region's hills.

FAR RIGHT: Champagne bottles are stored head down as the wine ages in natural limestone cellars. The bottles are cleaned, the sediment is expelled and the wine is topped up with a special dosage of sugars at the time of bottling.

RIGHT, AND BELOW FAR RIGHT AND CENTRE: Pink Champagne is becoming an increasingly important category in the market, and the choice on offer continues to grow.

RIGHT: The labels of a well-known Champagne, such as the sought-after Gosset, will be as familiar to the customer as any luxury brand logo. This brand recognition is an important element of the Champagne mystique.

OVERLEAF: A small part of the extensive cellars of a *Grande Maison* in the region. Many of these Champagne houses have up to 25 million bottles in store at any one time.

and this gives it a special nuttiness and depth, making it an excellent 'food wine' (see pp. 18–19). Red grapes thrive on the Montagne de Reims, grown in *communes* (administrative districts) with wonderfully evocative names, including Bouzy and Dizy. The wine of Bouzy is used to create pink Champagne, which is traditionally made by adding a dash of red wine to each bottle. This creates that lovely subtle tint we know and admire.

The fizz known as Blanc de Blancs, however, is made with 100 per cent white grapes, and this is the speciality of the Côte des Blancs, south of Epernay, where conditions are ideal for Chardonnay. Here you can find light, delicate wines from small producers called *récoltants-manipulants* (growers of grapes for Champagne who produce the wine from their own vineyards) on the one hand; and excellent value from the local co-operatives (such as the Union Champagne in Avize) on the other. It is interesting and enjoyable to compare the two, and to contrast them with better-known bottles from famous estates.

The early history of Champagne revolves around the mysterious figure of Dom Pérignon (*c.* 1638–1715), who is credited with perfecting the method of keeping those bubbles in, by wiring down the cork and reinforcing the bottle; he also made a significant contribution to the development of better-quality Champagne. Visit the Abbaye de Hautvillers, near Epernay, to learn more.

While in Champagne, do try the Marc de Champagne, a fiery clear spirit made with the grape residue. This local 'grappa' is the base for some marvellous sorbets and is also used to flavour liqueur chocolates shaped like Champagne corks, which make an excellent light-hearted souvenir from your travels – and are easier to carry than those heavy bottles of Champagne.

Fairs and Festivals

....

The New Year brings traditional celebration to many areas of France associated with Saint Vincent, the patron saint of wine, on 22 January (or the weekend nearest that date). Champagne is notable for its processions and feasting to mark this moment of the winemaking year; details are available from local tourist offices, such as: reims-tourisme.com; ot-epernay.fr

In Canada, the making of the unusual Icewine, pressed from grapes literally frozen on the vine, is celebrated with the three-week Niagara Icewine Festival this month: icewinefestival.com

Of course, January is summertime in the southern hemisphere, and in Australia the South Coast Wine Show in New South Wales showcases more than 160 wines from twenty-five wineries: southcoastwineshow.blogspot.com

LEFT: Even in the dead of winter, wine lovers flock to the Niagara Icewine Festival in Canada.

BELOW: Some of the vineyards of the house of Laurent-Perrier in Champagne.

BOTTOM: At the South Coast wine event in Australia, summer is a good time to assess wines from a previous vintage.

Stéphane Tsassis

CEO of Champagne Laurent-Perrier, France

· ·

Stéphane Tsassis worked for many years in the fragrance sector, including for such companies as Maison Guerlain. He became Chairman of the Management Board of Laurent-Perrier Champagne in 2009.

laurent-perrier.fr

What makes Laurent-Perrier stand out as a brand?

For more than sixty years Bernard de Nonancourt (owner, Laurent-Perrier Champagne) has dedicated his life to innovating and perfecting his Champagne. His forward-thinking approach has set the house of Laurent-Perrier apart ever since he took over the business in 1949. His idea was to build a 'personality' for each Champagne within the Laurent-Perrier portfolio. The result is that Laurent-Perrier wines are all pure, elegant and fresh – the unmistakable Laurent-Perrier signature style.

Through de Nonancourt's leadership, Laurent-Perrier has achieved many audacious breakthroughs that create distinction for the brand. One was the use of stainless-steel vats (instead of barrels) for the first fermentation, to extract the finest aromas from the grapes. Laurent-Perrier was the only brand to do this at the time. In 1959 the house introduced the first multi-vintage prestige cuvée, Grand Siècle by Laurent-Perrier, and in 1981 Laurent-Perrier was the first house to reintroduce the concept of *brut nature* Champagne [produced without the addition of *dosage*; see p. 21], with its Laurent-Perrier Ultra Brut. The

company also pioneered the rosé category for Champagnes by launching its iconic Cuvée Rosé.

Which of your Champagnes would you recommend as a special experience?

I would spontaneously say the Laurent-Perrier Brut, as it defines what Champagne is all about. It is a pleasurable wine that you want to get back to: one glass easily leads to another. There is no saturation. We like to say it is a *vin de plaisir* [easy-drinking wine]. But since you want me to recommend a Champagne as a special experience, then it has to be Laurent-Perrier's Cuvée Rosé. It is a demonstration of our *savoir faire* and of our audacity through creativity. Bernard de Nonancourt launched his now famous rosé Champagne in 1968, at a time when it was not at all fashionable. It showed his great independence of mind. He believed in a high-quality rosé Champagne made through a rigorous selection of grapes by hand, and his version has now set the benchmark for this type of Champagne. Its pink colour is vibrant and gorgeous, and the aromas of freshly picked red fruit hit you on the first sip. It goes well with Indian or Asian food, too.

Could you give me a very brief history of the company?

The Laurent-Perrier house has its origins in two families: those of Eugène Laurent and Mathilde Emilie Perrier. When they married, Eugène Laurent was a cellarmaster and Mathilde Emilie Perrier was a member of a farming family. Their names represent the two founding elements of the brand – the know-how of wine, combined with the know-how of the soil and nature. These remain the forces behind Laurent-Perrier. The present owner, Bernard de Nonancourt, took over the business from his mother, Marie-Louise Lanson de Nonancourt, who had purchased it from the Laurent-Perrier family in 1939. Bernard has continued to uphold the original principles of the firm's founders.

How do you see the future of Champagne in this competitive world?

The climate and *terroir* in the Champagne region are unlike those of any other winegrowing region in the world. These qualities, combined with the time-honoured skill and winemaking expertise developed over generations, are what make Champagne completely unique, and will hopefully ensure its enduring success.

Finally, what dishes would you serve with your Champagnes for a celebratory meal?

With our prestige Cuvée Grand Siècle, only the most refined ingredients will do: I would serve it with foie-gras canapés, lobster or langoustine as a starter, and a *Poulet de Bresse* with truffle and a simple purée of mashed potatoes as a main course. [These fine dishes are all classics of French cuisine and often served at formal dinners in France.]

February

All About Grape Varieties

February is a grim month for weather in most vineyard regions; of course, in the southern hemisphere it is the exciting time of harvest, but elsewhere it is a season of hard work, with pruning, as in January, a priority. As we have seen (pp. 13–14), pruning is critical for reducing yield so that the bunches of grapes that grow will be of intense flavour and concentration. There are many different styles of pruning, but all have the aim of improving the subsequent wine. From *gobelet* in Beaujolais to *Guyot* – a system of training vines, using canes – in Bordeaux, the vines look immaculate once they have been trimmed.

Of course, as vines' growing habits differ, each variety may need slightly different attention. In the great classic wine regions, there are certain grape types that are inseparably associated with the area. For instance, Bordeaux red wines are made predominantly with Cabernet Sauvignon, Merlot, Cabernet Franc, Petit Verdot and just a hint of Malbec. White wines are Sauvignon Blanc and Sémillon, be they rich, sweet Sauternes or ultra-dry Entre-Deux-Mers. These varieties have all migrated around the world to newer zones of production, but interestingly, some do much

OPPOSITE: The rolling hills of the Veneto, Italy, are snowbound at this time of year, but in the summer will be covered with lush growth of the Prosecco vine, the source of the elegant sparkling wine of the same name.

BELOW, CLOCKWISE FROM TOP LEFT: Prosecco grapes from the Veneto region of Italy; Cabernet Sauvignon grapes from Bordeaux, France; fine white grapes from Trentino, Italy; New York State grapes displaying uneven ripening owing to weather conditions during the growing season.

ABOVE AND ABOVE RIGHT:
These fine Burgundian Pinot
Noir and Chardonnay grapes
are the only varieties used for
wines produced in this region.

RIGHT: The Sauvignon Blanc
grape is virtually synonymous
with New Zealand wine, and
Montana is a towering force
in the industry there.

better than others. Malbec, for example, is now a star grape in Argentina, overshadowing even Cabernet Sauvignon; while Semillon has become an invaluable grape in Australia.

There is a definite hierarchy in grape varieties. Even in today's fast-changing and democratic wine world, certain vines enjoy celebrity status and are considered to be the aristocrats of the vine. The most highly prized red wines are made with Cabernet Sauvignon, Merlot, Pinot Noir, Nebbiolo, Sangiovese and Syrah grapes. White grape types that have this status include Chardonnay, Sauvignon Blanc and Riesling. One reason why these grapes attract such acclaim is tradition: Chardonnay and Pinot Noir between them produce Champagne, Chablis, and all the fine wines of Burgundy. Another reason they are sought after is that they are tricky to grow well. Merlot and Chardonnay will thrive virtually anywhere, although to make a true wine of quality they need close attention in vineyard and winery. By contrast, Pinot Noir is reluctant to ripen and can be almost undrinkable if not treated with skill.

Some of the finest grape varieties are now distinctly unfashionable; wine, like clothes, has its fashions. Our ancestors would have relished Gewürztraminer from

Alsace, Riesling from the Rheingau in Germany, and Tokay, made with the Furmint grape in Hungary and once a favourite of the Russian tsars. These wines still exist, and are as impressive as before, but they now rarely feature on wine lists. Some are too fragrant and seem sweet to our taste buds; others have labels that are so complex that the average wine lover simply becomes befuddled. Labelling is an absolutely fascinating and key feature of a grape's destiny. Not long ago, a wine merchant would list claret, hock, port and many other wines that made no mention of their origin. The actual grape varieties involved in the blend (most wines are blended, a decision made on grounds of quality) were not mentioned at all on the label. Even the merchant would be unlikely to know the precise mix of obscure Portuguese varieties in a vintage port; and in Châteauneuf-du-Pape, a hugely popular full red wine, there are no fewer than thirteen grape types legally authorized for use in the blend.

There has been a radical shift in wine labelling over the past twenty years. Wines from the so-called New World once called themselves 'Chablis', 'Burgundy' or even 'Dry Sauternes' to attract the buyer, but consumers have acquired more sophisticated understanding in the wake of what is known as varietal labelling, where the actual grapes used in making the wines are named on the label. A couple of decades ago, this was relatively unusual; although Alsace had always named varietals, and in California there was a trend for '100 per cent' wines, made with a single grape variety, most other blends did not carry details about their constituent elements on the label. These days, however, supermarkets sell dozens of Chardonnays and Cabernets from all over the wine world, and we ask for a glass of Merlot or Pinot Grigio, not claret or hock, in a wine bar.

LEFT : Varietal wine labelling has always been important in the California wine industry. This bottle of red wine displays the grape variety in letters as large as those used for the producer's name.

LEFT, BOTTOM: In Italy wine producers in the far northern regions, including Trentino, Alto Adige and the Veneto, were pioneers in varietal labelling. They also planted varieties originally associated with France, such as Merlot and, as in this instance, Chardonnay.

OVERLEAF: Grape pickers must work rapidly in order to maintain grape quality and avoid premature fermentation before the bunches reach the winery.

In the UK, Australian wines led the way in creating demand for varietal wines. Where once the winemakers jokily named their wines Kanga Rouge and Wallaby White, they soon collectively realized that making wine drinkers crave Chardonnay or Shiraz would pay greater dividends. This has led to an interesting shift in comprehension. The producer, shipper or brand was once all-important in choosing wine, and it is still a great help to know a few key names, especially in France, Italy and Spain: look out for familiar Louis Latour, Calvet or Antinori and you may feel reassured when confronting a smart wine list. But it is much easier just to ask for a varietal wine. If you love Merlot from Chile, you will not need to perform a memory feat next time you ask for a glass. If you crave Château Le Pin from Pomerol (a very expensive, collector's wine made with the same grape), however, you will need deep pockets and a good memory for detail.

In short, varietal labelling echoes our consumer society. We demand choice, and we also want to make quick decisions. Seeing a familiar grape type on the label reassures the buyer and helps to speed up the process. This may sound cynical, but since most wine is purchased to be consumed within three days or less, it makes great sense.

The great European regions, such as Burgundy and Tuscany, led the way in the concept of *terroir* – planting the right grape type in the right conditions and in the correct soil type. Now winemakers all over the world have learned their lessons and are confidently promoting their wines with varietal labels. Ironically, it is the very finest classic zones, such as Bordeaux, that have been slowest to realize how the world has changed, and now it may be too late for them to save many smaller vineyards. Faced with a choice between Australian Cabernet Sauvignon and Cru Bourgeois claret, most consumers will be reassured by the varietal name, which will prevail for much of the time. In this regard the Europeans need to look to their laurels and join the twenty-first century.

Romance and Wine

· ·

It is generally agreed that you should look directly into the eyes of your companion when clinking glasses. Otherwise, you may be unlucky in love!

We live in interesting times. On the one hand, it is generally acknowledged that delicious wine and food are the perfect prelude to romance; on the other, there is constant pressure to cut back our alcohol intake, watch our weight and feel generally paranoid about our health. Why not follow your heart for a change? February is the perfect month for diversion, especially in the northern hemisphere, as wintry weather forces us indoors and spring seems far off. It is no coincidence that Valentine's Day falls in February.

This is the month to get into the kitchen and the cellar and concoct an enticing feast for the one we love. Is wine an aphrodisiac? A biologist would solemnly agree that Champagne bubbles do hit the stomach lining with a jolt, and that the sudden release of tension we feel after a glass is at least partly down to physiology. But there is far more to seduction than just 'chemistry'. The rituals surrounding the preparation of food, and the serving of wine, all make a moment of theatre and even mystery: a great recipe for romance.

To make an impression, ditch the screwcap bottle and the unadorned kitchen table. Find a tablecloth, decant your wine into a stylish piece of glassware (this need not be a classic decanter; be inventive) and scatter some flower petals around and about. Candles are, of course, essential. Now for the choice of wine and food; it is important to strike the right note. Too much rich food

and wine, and your lover will be snoring by the time you serve dessert; too minimal, and that wonderful replete feeling will be missing.

On the following pages are some suggestions for menu combinations that could suit the occasion: choose luxury ingredients that are generally thought to possess seductive properties; go for a little chilli heat; or opt for simple, comfort food. All the suggestions are accompanied by ideas for complementary wines.

Casanova

~

Aperitif: Vintage Champagne or top-quality New Zealand sparkling wine,
such as Pelorus. Serve with oysters along with plenty of Tabasco and lemon,
and thinly sliced bread and butter.

Main course: Steak Tartare with a raw egg; poached asparagus.

Wine: St-Émilion, Pomerol or Barolo, previously decanted and left to warm slightly
by the open fire.

Cheese: Really ripe Époisses or Brie, served with walnut bread and more red wine.

Dessert: *Pots au chocolat* (chocolate mousse) with fresh figs.

To make the chocolate mousse, chill two cocktail glasses. Break four squares of rich dark chocolate into a double boiler, and warm until melted. Remove from the heat, and stir in a very small knob of butter and two egg yolks. Whisk the egg whites until firm, and fold gently into the chocolate mixture. Pour into the glasses and chill for a couple of hours.

Wine: Moscato di Pantelleria.

Hot Stuff

~

Aperitif: Bellini made with Prosecco and fresh, crushed white peaches (or use
the canned version). Serve with mixed nuts, tossed in a frying pan with
crushed black pepper, sea salt and a dash of Tabasco, and some smoked
mussels (from a can is fine) on cocktail sticks.

Main course: Fiery fish stew made with a mixture of langoustines, lobster, monkfish and
scallops, all in a hot, peppery sauce made with fish stock.

To start the cooking, fry some shallots in a large pan, then add a good dash of inexpensive brandy and flame briefly. This will add serious depth of flavour. Add the seafood, stock and some cayenne pepper or smoked paprika, and simmer gently for at least an hour. Serve in a gleaming copper bowl or Portuguese cataplana, and offer crusty bread and a green salad.

Wine: Douro red from Portugal, or a top-quality Australian Semillon.

Dessert: Ripe mango wedges and organic ice cream.

Wine: Gewürztraminer – scented and sweet, yet dry on the finish.

Retro

~

Aperitif: Kir Royale made with Champagne and *crème de cassis* or *crème de myrtilles*.

Starter: Scallops baked in their shells with a dash of Pernod, cream and a dusting of breadcrumbs, Parmesan and chopped parsley.

Wine: More Champagne or a white Burgundy.

Main course: Roast rack of lamb with dauphinoise potatoes baked with cream and lots of black pepper; buttery green beans and baby carrots.

Wine: Your favourite Cabernet Sauvignon.

Dessert: Ripe strawberries and a pot of chocolate for dipping, fondue style.

Wine: Muscat de Beaumes de Venise or Australian Muscat.

Of course, not every lovers' meal is dinner; how about breakfast or brunch? Champagne or Buck's Fizz (Mimosa if you're an American) is the default choice. Think a little differently and serve Black Velvet (Champagne and Guinness) with a good old fry-up, including Irish Champ (fried potatoes and cabbage – much more delicious than it sounds, although it must be cooked with the very best butter). Or how about a really delectable, chilled bottle of naturally low-alcohol Riesling from Germany, Austria or the New York Finger Lakes region, paired with the ultimate Continental breakfast: fine Parma ham, sliced Emmental and freshly made rye bread. Add some chilled grapes (peeling optional!) and your taste buds will be tingling.

The rich, complex flavour of Barolo, produced from the Nebbiolo grape in Italy's Piedmont region, benefits greatly from decanting, up to twenty-four hours before serving. It is interesting to note, in chilly February, that the variety's name derives from the Italian word for 'fog' and refers to the climate in these northerly vineyards.

Visiting the Vineyards of Veneto and Trentino

The map above shows the Italian regions of Veneto and Trentino. These areas are noted for several wines, including Soave, Valpolicella, Bardolino and Prosecco.

It is Mardi Gras this month, and from Rio to New Orleans there will be parties. In Europe, one of the greatest is the Venice *Carnevale*, where, as masks are donned, inhibitions are shed. Although there are no vines being cultivated in St Mark's Square (but there are vines within the city limits; see the interview with Gianluca Bisol, p. 41), there are vineyards just on the doorstep, and Venice can be the starting point for a marvellous exploration of the Veneto and Trentino regions of Italy.

A visit to the region begins with a flight to Venice, Verona or Treviso. From Verona, it is easy to take a tour of the vineyards of Soave and Valpolicella, as well as Bardolino. All of these are light, easy-going wines that can be most deceptive. At their least expensive, in cheap *trattorie* outside Italy, they are dull and bland, but sample them in their native habitat and they take on true *bella figura* – a phrase that summarizes the Italian obsession with making a good impression on the world, presenting a 'good (or beautiful) face'.

One excellent way to spend time in this region is to base yourself around Lake Garda, perhaps in Gardone, take in local lakeside sights and then divert to the vineyards. Be sure to visit Pieropan for Soave and Allegrini for Valpolicella if at all possible, to discover where the benchmark is set for these wines. Try a Recioto or Amarone version to see why Italians love these two wines and often feature them at weddings and other feast days. Recioto is made by semi-drying the grapes, then fermenting to dryness, yielding a higher alcohol content and a rich flavour. Amarone has even greater depth of flavour, and a touch of bitterness, beloved of Italians because it complements food cooked in oil or butter.

Drive north and you will find yourself in the mountains of the Trentino; fruit trees mingle with vines, and the climate is cooler, even in summer. A cool climate is advantageous for creating wines with a full flavour, and here there are Pinot Grigios that have real style and taste. There are also fine wines made with Merlot and other red varieties. Look out for wines from Cantina La Vis for reliable quality. Many wines in this area are made by co-operatives: harvesting on these slopes is hard work, and sharing machinery and manpower makes a lot of sense. Visiting and tasting at a co-operative is a really worthwhile experience, and offers quite a contrast to the slightly artificial romance of some more commercial cellars.

Speaking of romance, Verona has a plethora of attractions, not least Juliet's balcony, where she is said to have wooed, and been wooed by, Romeo. If you are thinking of romance, instead of Champagne, choose an Italian alternative from the region: in the Trentino, there are fine, Champagne-method sparkling wines made by Ferrari, and, from the Veneto, delicious, fragrant Prosecco. Heading east from Verona, do not miss the

RIGHT: Even the islands of the Venice lagoon grow grapes, and many are also fertile places to cultivate fruit and vegetables for the market in the city.

BELOW: Making an entry to Venice's Grand Canal by water taxi is the best way to sense the history and grandeur of this extraordinary place.

BOTTOM: Venetians gather in small boats for the Redentore festival held on the third weekend of July. The festival originated as a gesture of thanks for survival of an outbreak of plague in the Middle Ages.

charms of Padua, a historic university city, and Vicenza, often overlooked in the rush to Venice. Drive on to Treviso and you are at the gateway to the utterly charming Prosecco region, in a part of the Veneto that is definitely off the beaten tourist track. Conegliano is the centre of this production area, set amid rolling green hills reminiscent of those in a Renaissance painting.

Each hill is literally covered with grape vines, all of a single variety: Prosecco. Not long ago, the light sparkling wine of this name was seen as a strictly local charmer, but now it is invading the bars and restaurants of the world. Its accessibility in terms of both taste and price when compared to Champagne make it very popular with younger drinkers. As Prosecco is produced by the Charmat method (in unglamorous tanks rather than individual bottles), the bubbles are bigger and softer than the fierce fizz of Champagne. Prosecco, with its honeysuckle aromas, lends itself to such cocktails as the classic Bellini, made with white peaches, and a favourite in Harry's Bar in Venice; also the knockout Spritz, a Venetian after-work special that involves Campari or other bitters mixed with Prosecco, sugar syrup and occasionally other mystery ingredients, all of them delectable.

RIGHT: The green countryside of the Veneto is only a short drive from the magnificent cities of Venice and Verona.

FAR RIGHT: The mountainous Trentino region has many natural caves that over the generations have been turned into wine cellars.

BELOW: In the Trentino Alps wines are made with a great variety of grape types, ranging from familiar Pinot Grigio and Cabernet to lesser known white Nosiola, and red Lagrein and Marzemino.

North of Treviso there is more to explore for the wine lover in the red-grape zones where excellent Merlot and Cabernet Sauvignon are produced. Again, the landscape features vine-covered slopes and, indeed, this is Italy's second-largest production area (after Emilia-Romagna, home of Lambrusco). Forget the simple Veneto Merlot you know at home; to see how good these wines can be, visit a small producer, such as Località Le Ragose in Arbizzano or the elegant La Montecchia wine resort near Padua. Be sure to take a phrase book as, once you leave Venice and Verona behind, English may not be on the menu.

Left: A very modern-day seller of bird seed makes a slightly incongruous figure against the Byzantine pomp of the Basilica San Marco, Venice.

Above: The boundaries of Bisol estates, in the Veneto region, are marked by trees.

Below, left and right: Bisol's fine Prosecco wines are produced in a variety of styles, including a pink version; the Jeio selection is named after a Bisol family member's nickname.

Fairs and Festivals

....

There are many places to relish the hedonistic delights of Mardi Gras: Venice, of course, but also New Orleans, or Germany, where the celebration is known as Fasching and, as with similar festivals elsewhere, affords a licence to behave outrageously for just one day.

To find other February wine fairs, follow the New Zealanders, who know how to make the most of their vintage season. Go to Marlborough for the lovely Sauvignon Blanc: wine-marlborough-festival.co.nz

You could head to Oregon, in north-western USA, for their Truffle Festival. Truffles, a culinary delicacy normally associated with France or Italy, thrive in the Pacific Northwest too, and Oregon has some exceptional Pinot Noir wines to accompany them: oregontrufflefestival.com.

At the end of February, visit the lovely vineyards of the Loire Valley in France, so romantic in winter mist, for the Anjou Wine Festival: fetedesvins-anjou.fr.

Right: New Orleans, famed for its Mardi Gras celebrations, also has wine producers, such as Pontchartrain Vineyards.

Below: Venice's colourful *Carnevale* tradition, which had lapsed in the nineteenth century, was rejuvenated in the late 1970s.

Bottom: The summer sun of New Zealand in February makes the Marlborough Wine Festival a great way to celebrate the harvest.

Gianluca Bisol

General Manager of Bisol Desiderio & Figli, Italy

· ·

Bisol Desiderio & Figli is a family-run, highly successful winery near Treviso. Bisol is well known for its superb Prosecco.

bisol.it

Is Prosecco a rival to Champagne, or is it in a different category?

If Champagne represents the king of sparkling wines, I imagine Prosecco, for its distinct elegance, is the prince of bubbles.

Is it a 'young' drink?

For its freshness and versatility, Prosecco is well loved by young people: furthermore, it has a good quality–price ratio and matches perfectly with many dishes, including Japanese and fusion cuisine. It is a modern wine.

Tell me, what makes Prosecco unique?

There is no other sparkling wine with the same characteristics as Prosecco. 'Prosecco' was one of the hundred newly added words that appeared in the 2008 edition of Merriam-Webster's *Collegiate Dictionary*. Authentic Prosecco is produced in a well-defined area in north-eastern Italy known as Conegliano-Valdobbiadene, by producers who comply with strict regulations governing the growing area, grape content and production method.

How about the other wines from the Veneto? Which are your favourites?

I love the wines from the Veneto hills, such as the Raboso Piave, a full-bodied wine with a lot of character. I am fascinated by Venetian wines. I find the Dorona variety very interesting: it is a white grape, typical of the Venetian lagoon, that I love for its flavour and minerality. I discovered that this ancient grape, once grown in the lagoon, was nearly extinct, and after careful research and recultivation, we are replanting an ancient vineyard on the island of Mazzorbo, in the Venissa estate.

What about the future for Prosecco: where do you go from here?

With the transformation of Indicazione Geografica Tipica (IGT) Prosecco into Denominazione di Origine Controllata (DOC), and of Conegliano-Valdobbiadene from DOC into Denominazione di Origine Controllata Garantita (DOCG), the perception that the market will have of Prosecco will be clearer. [Within Italian wine regulations, DOCG is a superior category to DOC; IGT signifies wines that are typical of a particular region, and good value for money.]

Could you give me a little Bisol history, please?

The history of what many consider to be the 'first family' of Prosecco is a long and distinguished one. By all accounts, it begins back in 1542; however, the modern winemaking history of Bisol really begins with the brilliant work and great passion of Eliseo Bisol (1855–1923), credited with the rise of the estate, which in many respects is a true jewel of Italian oenology. It is not by chance that the estate owns more than 125 hectares (308 acres) in some of the choicest sites of the gently rolling hills around the little town of Valdobbiadene.

Tell me about your corner of the Veneto region and why it would be attractive to the wine tourist.

The tower of Conegliano's castle is where the *strada del Prosecco*, 'Prosecco Road', really begins. The route was outlined in 1966 and runs for about 45 km (30 miles) through the province of Treviso, among the hills of Conegliano, Feletto, Quartier del Piave and Valdobbiadene. Strategically located between Venice to the south and Cortina to the north, the Bisol estate is situated in a beautiful landscape, through which, from gentle foothills to the steeper slopes, the *strada del Prosecco* winds its way through the vines. Superb local restaurants offer an opportunity to taste the excellent wines of the region matched with traditional and modern cuisine.

March

Wine and *Terroir*

....

OPPOSITE: The Faustino winery
in La Rioja, Spain, uses a vast
array of American oak barrels for
ageing its fine red wines.

BELOW: A worker trims back
excessive growth on vines in
La Rioja to ensure the grapes have
the best opportunity to ripen and
to produce quality wine.

For the wine novice, there is something rather terrifying about the French insistence on the importance of what their *vignerons* call *terroir*. It is the magic ingredient that adds mystery to the winemaking equation, a reason for the French to assert national superiority even in the face of economic odds. Confidently proclaiming their wines to be the best in the world, they will cite their intimate understanding of *terroir* as their advantage.

So what exactly is *terroir*? It could be translated simply as 'terrain', the earth in which the vine grows, and the rocks beneath what is often quite a thin layer of soil in the vineyard. In the southern hemisphere, March is the time of harvest, but in the northern hemisphere, ploughing is a very important activity at this time, to 'turn in' (plough in) competitive plants between the rows and allow the development of potential grapes to begin in each vine. 'Green manure' (see p. 14) may also be used – ploughing in such plants as rye or mustard to add natural fertilizer to the soil. Alternatively, branded or organic fertilizers are added, in judicious amounts.

However, the addition of too much fertilizer to a vineyard actually takes away from the whole meaning of *terroir*. A vine likes to have a battle for existence; some of the finest wines are produced in austere conditions that would be hopeless for other crops. The giant pebbles of the Rhône Valley, the steep slopes of the Mosel in Germany and the harsh, dusty terraces of the Douro in Portugal are all European examples of unlikely settings for successful agriculture. Yet the vines here are the source of magnificent wine: Châteauneuf-du-Pape and Hermitage, fine white Riesling, and classic port respectively.

Soil components and the geological make-up beneath undoubtedly influence the eventual wine, if the winemaker is willing to play along and keep the wine true to its origins. Certain clichés appear to apply to fine wine: for example, there is a preponderance of good wine made on limestone outcrops, such as Chablis, Champagne and Sancerre. The soil and its mineral components are only part of the story of *terroir*, however. There has been intensive study applied to the subject, nowhere more passionately than at the University of Bordeaux. Here, in the heart of the region that produces some of the world's most sought-after wine, academics have analysed and discussed the soils in top vineyards for years.

As a result, there is now general agreement that *terroir* is not simply the sum of the soil and its tending. There are other, vital elements involved, including the amount of sunshine and rain the grapes receive in a season, the correct choice of grape variety, and the angle of the vineyard and its natural drainage. Far from being a complete mystery, *terroir* can be partially explained in terms of these various indicators.

The choice of site for vineyards that have long yielded fine wine (such as those in the Côte d'Or region of Burgundy) was evidently made with these factors in mind – perhaps not quantified scientifically, but nonetheless based on logic. Free-draining slopes allow the vine to thrive, and there is less danger of rot and mould; rain and sun in proportion allow the grapes to mature steadily, which is vital to a successful harvest. If grapes are grown on a very fertile plain, they are certainly vigorous and grow rapidly, but much of the plant is excess leaves, which overshadow the fruit and make the

ABOVE: Grapes are protected from greedy birds just prior to the harvest at the Torres vineyards in Catalonia.

LEFT: The Nahe region of Germany is a good example of rocky growing conditions; this lends character to the wine, and the sun-warmed stones also release heat at night.

RIGHT: A winegrower holds a handful of Terra Rossa soil, a type of clay derived from crumbled limestone. Especially associated with the Coonawarra region of Australia, it is also found in La Mancha, Spain.

FAR RIGHT: These serried ranks of vines in the Pyrenean foothills of Catalonia demonstrate how a good *terroir* can be planted to make use of every inch of valuable wine-production land.

wine taste grassy and unpleasant. The bunches may also receive too much water, as it settles on the rich soils, and this will make them swell, thus diluting the components that flavour the eventual wine.

Matching a grape variety to a *terroir* is a fine art, and this is where the New World has been at some disadvantage. During the past thirty years, the concept of *terroir* was largely derided in the United States, South America and Australia. What was seen as important were the hours of sunlight, the vigour and health of the vine, and the work in the winery to make good, clean wines. Modern technical skill can bring together all kinds of grapes from different origins and make them taste pleasant enough.

Yet these fruity, commercial wine styles often lack the character and depth found in classic European wine. Some of the fault may lie in growing grapes in the wrong locations: in a hot climate, Chardonnay becomes flowery, rich and oily on the palate, while Pinot Noir tastes raisiny and dull. In recent years, as wine consumers learn more and demand more from the winemaker, there has been a strong trend among producers to return to the land and respect the concept of *terroir*. The actual word may not be

used, but the emphasis on the work in the vineyard and the importance of regional or even single vineyard character is a gesture of respect to this idea.

Even alcohol levels have a link to *terroir*. If grapes are allowed to ripen rapidly in full sun, and then are fermented out to the limit, the result can be a wine that has initial appeal, thanks to heavy fruit and the roundness of the alcohol on the palate. Those 'legs' of viscosity that run down your wineglass are a clear indication of this. Yet these wines have no staying power and may lose their 'personality' in a matter of months. They are not, ultimately, impressive.

If grapes are grown in the right zone, ideally on well-drained slopes, tended carefully to allow just enough sunshine on the bunches, checked for health and sprayed if necessary, then harvested in peak condition, they have every chance of becoming a good or excellent wine. The eternal mystery of *terroir* is that even where all these factors are in place, the wine may be sound and agreeable, but not impressive. There is still more to learn about the link between a wine and its origins in the earth, which is one reason why *terroir* is such a fascinating and evolving subject.

Hearty Eating

···

In northern Europe, March is a month of false starts. The weather may turn unseasonably mild and fool the vine into growth (a dangerous moment), or it may be as hostile as midwinter. After so many months of chill, warming dishes are always welcome, and there are some ideal wines to accompany them. This is the time of year to plan some leisurely Sunday lunch parties, perhaps followed by a good long walk to aid digestion after eating. So, with this plan in mind, indulge a little and think in terms of wines with a relatively higher alcohol content and plenty of full flavour.

In anticipation of the guests' arrival, this could be a rare occasion for decanting wine. There are two or three possible reasons for doing this: to air a wine that is fairly heavy and alcoholic, encouraging it to open out and reduce that sudden 'hit' of alcoholic strength; or you might want to surprise your guests with an unusual wine, served in a carafe rather than from the bottle. Finally, you may have a fine or rare wine that is just at the right point for drinking, but needs decanting to check it and ensure that there is no sediment to surprise the palate. Make the serving of the wine something of a ceremony.

Now to think about suitable menus. It is not essential to spend huge sums on a hearty lunch: the key is taking time to prepare and then cook at a leisurely pace. Why not make dessert the day before, if possible? Or, if you are making a soup to start, that would almost certainly benefit from early preparation; soups seem to 'marinate' in the fridge and gain more flavour.

A sturdy stew of beans and sausage is complemented by the spicy, nutty flavours of a quality Amontillado sherry from Jerez, Spain.

A Serious Sunday Lunch –
with a Spanish Twist

∾

As an aperitif, serve some well-chilled fino sherry and a small selection of Spanish-inspired tapas, such as Serrano ham on sourdough bread; anchovy-stuffed olives; and sweet pimento peppers.

Continue with the sherry and an unusual soup: Crème de la Vierge. This is made with small, sweet and peppery turnips, blended with good chicken stock, onion and potato and a stir of rich cream to serve. Add some chopped chervil if available, and you will find that a turnip has never tasted better.

For the main course, make a dish of Spanish *alubias*, a version of French cassoulet made with spiced sausage, chorizo, duck and other game, such as rabbit. Mix these according to availability but be sure to add the spicy sausage for depth of flavour. Include white beans, plenty of tomatoes and some spicy pimentos. The *alubias* can be made over several days and reheated a couple of times before it is finally served, adding more stock and some red wine to taste. You may also want to fry some breadcrumbs in butter and add them on the top before finally baking and serving in a large earthenware dish. To accompany, offer some healthy greens, such as curly kale, savoy cabbage or chard, steamed with a knob of butter. Add rustic bread to mop up the delicious juices.
Serve this with a full-blooded Rioja Reserva with a few years' bottle age. Alternatively, think New World and consider a Merlot from Chile, a Malbec from Argentina or an Australian Shiraz.

Follow with some cheese: continue with the red wine and offer Spanish Manchego and quince paste (*dulce de membrillo*, available from good cheese shops or Spanish importers, such as Brindisa in London), or some top-notch English farmhouse Cheddar.

For dessert, make a simple fruit compote by gently simmering a selection of dried and fresh fruits, such as apricots, figs, prunes and oranges in a sugar syrup flavoured with star anise and a few pods of cardamom. Serve with fresh custard and some Moscatel de Setubal from Portugal or a sweet sherry, such as Oloroso. Accompany with orange-flower shortbread – just add a good dash of orange-flower water to a standard shortbread recipe – to make it all a perfect mix.

From an Italian Farmhouse

‿

This menu summons up those relaxed meals often found on Sundays
in Italian *agriturismi* – a wonderful combination of farm and holiday
accommodation with a family atmosphere.

For the aperitif, offer a fresh glass of a Franciacorta, a sparkling wine from northern
Italy, served with bruschetta made with chopped tomato, chopped black olive and
good olive oil, all well seasoned.

Follow this with a dish of fresh egg noodles served with a rich *ragù* – if you want a touch
of authenticity, this should be made with hare or wild boar – and some fine Chianti or
Vino Nobile di Montepulciano (not to be confused with the grape of that name, which
may yield some fairly unexciting wines).

For the main course, roast an organic chicken and serve with side dishes of polenta,
oven-baked squash and green beans. Pour more red wine or, for those who prefer white,
a zingy Gavi di Gavi.

Finally, freshly made panna cotta is a simple, light and creamy dessert, which could
be served with a sauce made from berries, or just some rustic honey. Offer *cantucci*
biscuits and a glass of Vin Santo, a spicy dessert wine made with grapes that have
been air-dried before pressing.

Visiting the Vineyards of Spain

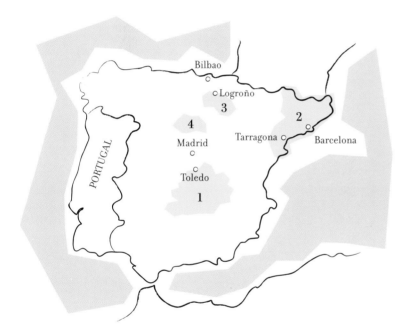

1 La Mancha

2 Catalonia

3 La Rioja

4 Ribera del Duero

The map above shows the wine areas of Spain that are discussed in this feature (see key, above). The numbering follows the order in which the areas are covered.

This vast and varied country has more land under vine than anywhere else in the world, so the choice of wine-tasting destination is limitless. There are rustic country wines made everywhere, often from apparently ancient vines that crouch on hillsides like malevolent sprites. On the other hand, there are also astonishing, futuristic wineries that rise from the plains of La Mancha like motherships in a movie. Technology has definitely arrived here, and it has improved overall quality in the average Spanish bottle to the point where wine connoisseurs and enthusiasts are warm in their praise.

It is difficult to decide which areas hold most excitement for the wine lover. From a logistical point of view, the wines of Catalonia are very accessible. Flights to Barcelona abound, and this city is stylish and youthful, and filled with magnificent art and architecture: think of Antonio Gaudí (1852–1926), architect of the amazing cathedral of the Sagrada Familia (scheduled for completion in 2026), and of artists Picasso, Miró and Dalí. There are also excellent restaurants featuring modern cuisine prepared by imaginative chefs. So, begin your journey here and travel beyond the city to visit the impressive and extensive vineyards of wine producer Torres, the standard-bearer for Catalonia and still a source of reliable and elegant wines of all styles and colours. Miguel Torres made the public believe in Catalan wine; at the same time, Cava, that affordable and attractive fizz, the vast majority of which is produced in Catalonia, has charmed countless drinkers worldwide. Not many of them realize, perhaps, that this wine is still produced by the complex traditional method (as in Champagne; see 'January', pp. 20–25) and that a

visit to the cellars of any large producer is a truly memorable day out. The vast limestone caves used as cellars have an almost ecclesiastical air, and such producers as Codorníu and Freixenet know how to make their many visitors feel welcome.

Much less easy to love, but full of characteristic Catalan verve and passion, are the wines made in Priorat, inland from Tarragona; a high enclave of serious red winemaking that has been established on the international scene for only thirty years. The wines made in and around the oddly named village of Gratallops are produced from mature Garnacha and Cariñena vines with just a hint of Cabernet Sauvignon, Merlot and Syrah, to yield long-lived and genuinely unique bottles that are now the target of collectors.

Move on from coastal Catalonia and head inland across the province of Aragon to Old Castile and La Rioja. This extensive, celebrated region extends for some 120 km (75 miles) and encompasses quite a wide variety of styles and microclimates. Rioja Alta is generally viewed as the quality area, although good wines may also

come from Rioja Baja and Rioja Alavesa, over in Basque country. Most of the vineyards are sheltered from the harsh Atlantic gales by the steep Sierra de Cantabria, a forbidding wall of rock that rises above the dusty plains. Look for stylish and well organized modern wineries with such famous names as Marqués de Riscal, Faustino (see Interview, p. 57), Marqués de Cáceres or Martinez Bujanda, and taste some of the excellent red and white Riojas they produce.

Once, the wines here focused on lengthy oak age (often in American oak), but now most are far more accessible and adaptable to modern cuisine tastes. As for local eating, there is a fondness for simple grilled meats: suckling pig or roast kid are favourite showpieces, and the native white asparagus marries surprisingly well with the white wines, made from the Viura grape. Reds are produced from Tempranillo, a grape of considerable charm and fruit, which is very drinkable when young, but which also, in the hands of the right winemaker, has good ageing potential. When touring the Rioja region, be sure to visit and if possible stay in Logroño, heart of the

OPPOSITE, LEFT: The Spanish pioneered futuristic wineries, such as this 'mothership' style in La Rioja, designed by Jesús Marino Pascual as the Darien winery headquarters and opened in 2007.

OPPOSITE, TOP AND BOTTOM RIGHT: La Rioja's Ebro valley floor contrasts with the Cantabrian mountain range. The old town of Logroño and its cathedral are at the heart of the region.

ABOVE: Oak barrels still play a part in the making of both red and white Rioja, but there are also younger, fresher styles for immediate drinking.

BELOW: The bottle shape for Rioja, red and white, is not standardized, but there is a certificate of authenticity across the neck of the bottle to reassure the purchaser that this is the authentic article.

The remarkable Ysios winery is located in Laguardia, just to the north of the classic Rioja region. Its owners, Domecq, commissioned it to harmonize with the surrounding Cantabrian mountain range. Designed by Santiago Calatrava, the winery was inaugurated in 2001.

BELOW: Pink wines have been a success story in all markets across the world. The rosé wines of Spain, produced from such grape types as the classic Tempranillo, have an intense, cherryish colour and a full flavour.

RIGHT: This dramatic shot of the Rioja vineyards shows the arid growing conditions in the region.

winemaking country, with a fine cathedral, ancient city walls and an atmospheric old quarter.

Drive on for some 96 km (60 miles) south-west, and cross the slender Duero river, which will become the mighty Douro once it reaches Portugal and port country. In Ribera del Duero, vineyards are set high and the weather conditions can be harsh, with frequent late-spring frosts and late harvests as producers wait for full ripeness from their Tempranillo. This is red-wine territory, and, like Priorat, its wine was almost unknown until recent decades, but is now eagerly sought by restaurants and wine lovers worldwide. Vega Sicilia was the first pioneer wine to be made here, and it gained legendary status, before being joined by dozens more wineries, many of which now produce on a large scale. The high reputation of these wines has meant that many producers already successful in La Rioja or other regions are acquiring land here, and the fields of sugar beet that once proliferated are now valuable vineyards. It has been quite a gold rush for Spanish wine, and the country is an intriguing place to visit, so go and judge these powerful, characterful wines for yourself.

ABOVE: This well-ordered cellar at Faustino in La Rioja contains reminders of how wines were made in the mid-twentieth century, before technology revolutionized this region.

ABOVE RIGHT: A Torres employee in Catalonia collects grapes at harvest time to test for ripeness.

RIGHT: The traditionally styled Campillo winery sits at the foot of the Cantabrian mountains.

Fairs and Festivals

March is a lovely month to visit France's Loire Valley, which is misty and atmospheric at this time of year. In Bourgueil, home of excellent red wines made with the Cabernet Franc grape, the Fête des Vins de Bourgueil is a main event; it takes place in the town of Tours: vinsbourgueil.com

In the United States, the icy chill of wintry Long Island is mitigated by the cool sounds of the Winterfest, which adds a dash of music to the tastings of locally produced fine wines: liwinterfest.com

Meanwhile, in Australia late summer brings the Melbourne Food and Wine Festival, featuring culinary stars from around the world as well as wine tastings, and the Adelaide Hills Crush Festival, with more than forty participating wineries: melbournefoodandwine.com.au; crushfestival.com.au

LEFT AND BELOW: The opportunity to taste wine and enjoy relaxed musical events are combined at the Long Island Winterfest (left) and the Adelaide Hills Crush Festival in southern Australia (below).

BOTTOM: Melbourne's annual Food and Wine Festival celebrates both wine and gastronomy.

Antonio Pinilla

Export Director, Faustino, Spain

The Faustino company has grown, under family supervision, from its origins in 1861 to a global brand. Its wines are found on restaurant wine lists and in wine shops and supermarkets worldwide.

bodegasfaustino.es

Could you give me a brief introduction to the company?

The Faustino Group, a family company in its fourth generation, has become an unquestionable leader in making and exporting top-of-the-range wines. The winery has 650 hectares (1606 acres) of vineyards in the heart of the Rioja Alavesa region, distributed between the towns of Laguardia and Oyón. The excellent qualities of the clay-calcareous soil in these winegrowing areas, together with an ideal microclimate, afford the grape varieties that grow in the vineyards (Tempranillo, Graciano, Mazuelo and Viura) enhanced quality and unbeatable added value.

The winery's premises are home to more than nine million bottles, which are placed on the market only when Faustino's expert oenologists consider it appropriate. Approximately 51 per cent of the group's sales currently take place on the domestic market, and exports represent 49 per cent of the total figure, which makes the group the leading exporter of Rioja Gran Reserva wine in the world.

How would you define the Faustino Rioja style?

Faustino has remained true to its high quality and elegant style; we start to 'nurse' our wines in the vineyards, and we are constantly introducing improvements in our vinification facilities and winemaking techniques to enhance the quality of our wines.

How do you feel the wines of Rioja are perceived in the world today?

There are many other wines coming from different regions of Spain, but La Rioja continues to be perceived as the best red-wine region. The classic-style Rioja is considered by consumers worldwide to be incredible value – top-quality and aged wines at very reasonable prices – and the new generation of contemporary Riojas is also attracting international attention.

Tell me about the attractions of your region for wine tourism.

The Rioja area and wine are absolutely 'one soul'. Wine is a reference point for our culture in our region. Besides visiting any of our festivals, I strongly recommend taking 'The Way of St James' [the pilgrimage route to Santiago de Compostela], where you will be surrounded by vineyards and impressed with the traditional and modern wineries; take the time to stop in any medieval town to enjoy our gastronomy and wine, or learn more about the culture of wine and La Rioja in many of the different museums that you will find along the route.

How about the other wines of Faustino, in other regions?

Since it began making wines in 1861, the Faustino Group has grown, and it now has seven wineries in four of Spain's leading designations of origin. Most of its winemaking activity takes place in the Rioja Qualified Designation of Origin, an area that includes the Bodegas Faustino, Bodegas Campillo and Marqués de Vitoria wineries. In the Mancha Designation of Origin, the Group has Bodegas Leganza and, in the Navarra Designation of Origin, Bodegas Valcarlos. The Ribera del Duero Designation of Origin has witnessed the launch of one of the Group's most dynamic projects with Bodegas Portia. The Faustino Group also owns Bodegas Victorianas, the wines of which are made with grapes selected from a variety of winemaking areas in Spain.

The new generation of Bodegas Faustino has successfully upheld the family tradition of purchasing land as a guarantee of the quality of the wines made at its wineries; so much so, that Bodegas Faustino has become the leading vineyard owner in the Rioja region.

Finally, could you choose just one wine from your range and tell me why this is your top choice?

I will choose Faustino I, definitely an icon in Spain. Made predominantly with Tempranillo, balanced with Graciano and Mazuelo grapes, and aged in French and American oak, this is a very aromatic and complex wine, powerful yet velvety, with a long and elegant finish.

April

Organic and Biodynamic Wine

....

While in the southern hemisphere the vines in April are displaying autumnal colours, in the northern hemisphere this is usually the time when the vines are looking their neatest. After the ploughing in March comes the trimming of vine suckers (green shoots that take some of the energy of the vine away from the main fruiting stems) in April, and the result can be an immaculately tidy appearance. It could be argued that the focus on perfect order is excessive; certainly many organic and biodynamic winegrowers like to see wild flowers blooming and songbirds singing amid the vines.

This is a sure sign that their methods are working and bringing true biodiversity to the land.

There are far more organically grown grapes produced than there are organic wines. This may seem strange, yet once all the facts about organic production are known, it is all quite logical. The craft of the winemaker does not necessarily call for devoted attention to the land and the use of organic principles. If the winemaker starts with healthy, 'happy' grapes, it is a huge bonus, but then to make fully organic wine to certification standards usually means relinquishing the use of sulphur

OPPOSITE: The Vinedo Los Lingues vineyard, in the Colchagua Valley of Chile. The country's exceptional climate means that cultivation of organic vines is often easier here than in other world wine regions.

RIGHT: The lush coastal vineyards of Margaret River in Western Australia have been noted for fine Cabernet Sauvignon, Chardonnay and Semillon since the 1960s.

dioxide as well as of filtering and fining agents. In order to make a commercial wine that will survive transit in bottle to international markets, one has to ensure first and foremost that the wine will be deliciously drinkable when it reaches the consumer. Skipping the sulphur and so on can often lead to the wine becoming unstable, and, at the very worst, it will actually start refermenting in transit or arrive tasting oxidized or 'corked'.

This is every wine producer's worst nightmare, so why do so many passionately persist in making organic wine, all over the world? There is a deep core of belief among these people that the vine and its *terroir* deserve respect, and that using chemicals as short cuts to increasing yield is tantamount to cheating. So they avoid fertilizers, which can cause the vine to overproduce inferior grapes; fungicides, which may linger in the wine; pesticides, many of which are controversial; and also any hint of genetic modification (GM).

Of course, not every commercial grower is a villain, armed with vats of powerful products for dosing the vines. The great majority today work to a system the French entitle *lutte raisonnée* ('reasoned struggle'), meaning that use of sprays is kept to the minimum and the vine is allowed to find its own strength through the use of natural fertilizer, such as compost. This style of management is taken further by the organic growers, who also plant such cover crops as mustard or green rye between the vines, in order to help with nutrition and encourage deeper roots, meaning that less water is needed to support the vine. Then there are the biodynamic believers, who do spray their vines, but with very different substances from the usual choices. They use homeopathic quantities of silica or cow manure well diluted in water. There is just one constant: almost everyone in the wine world agrees that it is difficult to dispense entirely with a light spray of copper sulphate to fend off mildew.

The beautiful 'pincushion' flowers of protea, the national flower of South Africa, adorn this vineyard in the Western Cape; the region's extraordinary biodiversity has led to it being named a UNESCO World Heritage Site.

RIGHT: Organic vineyards tend to appear rather less manicured than these clean-cut rows of vines ready for machine harvest.

RIGHT, CENTRE: Colourful cover crops of mustard and grasses are a characteristic of many California vineyards; they are frequently ploughed in to enrich dry soil.

RIGHT, BOTTOM: Biodynamic winemakers fill cow horns with manure and quartz silica and bury them among the vines over winter to enrich the soil.

Only about 2 per cent of the world's wines are actually certified organic, but there are many more that are made with organic grapes, particularly in dry, consistent climates, such as those of Chile, Argentina, Spain and the South of France. The largest organic producer of all is Fetzer, in northern California, with more than 300 hectares (740 acres) of organic vines. Without sudden falls in temperature or unexpected rainstorms, it is far easier to watch over the rows of vines and avoid disease by natural means. One way is simply to plant a rose bush or fruit tree at the end of each row; if it shows signs of mould, then action will be needed, using a non-invasive spray. It is a huge feat to achieve organic status in the marginal climate of the Mosel, Champagne or Chablis; trying to make organic wine in the UK would be virtual economic suicide, as it has such changeable weather.

Talking of toughness, albeit of a more literal kind, red grapes often have thicker, more resistant skins (packed with those flavonoids that are said to be so good for us) than whites; as a result, there are fewer white organic wines. Despite this, some of the finest regions for organic production are in areas that are famed for white wine. The Loire Valley is home to many bold, experimental characters who create tremendous wines:

Above: Hand cultivation and picking are keynotes of the organic approach to winemaking.

Left: Encouraging biodiversity in the vineyards is a happy consequence of growing grapes organically or biodynamically.

these producers include such pioneers of the biodynamic method as Joly and Huet. In Burgundy, there are the houses of Leflaive and Leroy, the immensely expensive wines of which are made entirely biodynamically. In the South of France, Chapoutier is committed to the idea, and in the Ribera del Duero region of Spain, pricy Pingus is equally passionate about this concept.

Biodynamism all began with the ideas of Austrian philosopher and social thinker Rudolf Steiner (1861–1925), who linked the phases of the moon and the turning of our planet to the planting and harvesting of crops. This idea is as old as agriculture itself, and derives from the belief that our world is a living organism. Great wine comes from healthy soil, and to achieve this, organic methods are used, with adherence to a special calendar for various vineyard activities. The soil is enriched with tiny amounts of concentrated manure and natural quartz silica; these are put into cow horns, which are buried among the vines and left during winter, and then sprayed on to the vines in spring. At this point in the explanation, many shake their heads in disbelief, but the fact remains that many of the wines made under this regimen are the finest in the world.

At a time when green issues are top of the agenda all over the world, it seems only natural to show a little respect to vines. Replacing vast, bleak, machine-harvested rows with idyllic fields of wild flowers and healthy vines may seem romantic, but chemicals are not cheap, not good for us or the environment, and ultimately wasteful of resources. If avoiding excessive use of them means employing a few extra people in the vineyard, that may be a more sustainable way in which to proceed for the future. After all, if even famed Château Latour, one of Bordeaux's top five wines, is moving towards the organic ideal, this must be an idea worth taking seriously.

Matching Wine with Tricky Dishes

Pairing food and wine is an imprecise art. When it comes to decisions about what to serve with which dish, personal taste plays a very significant role. There is also the consideration that some of the ingredients used may come from countries and cuisines where wine is almost unknown. Now that people in Thailand, India and China are starting to grow vines and drink wine, there is a move towards making a match for the food from these countries, but it is still very early in their wine history to set any rules.

There are certainly some tricky elements of a meal that might challenge the classic flavours and styles of wine. Chilli peppers may numb the palate; lemongrass or curry leaves add a pungency that may overpower a wine; sumac and tamarind have a smoky richness that is very dominant. Beyond this there are various classic foodstuffs that are seen as unfriendly to wine, such as asparagus, artichokes, vinaigrette and chocolate (see p. 65).

On the following pages are a few guidelines that may help you to find a balanced combination of wine and food, whether you are cooking at home or ordering in a restaurant. Thinking of wine by style is a helpful starting point (see 'June', pp. 89–91). A difficult ingredient

Chablis and other dry white wines will complement rich, fatty dishes, such as terrine.

can be tackled in one of two key ways: by seeking a complementary wine that brings out its unique flavour, or by selecting a wine that will bring its own 'personality' to the table and create an interesting fusion of flavour. Another, rather decadent option is to go for a sparkling wine throughout the meal; the zesty fizz will enhance any dish, and even if the aroma and fruit in the wine are hidden at times, the bubbles are certain to charm.

Indian Cuisine

～

The predominant impression here is of richness and oiliness on the palate, especially when dishes are made with ghee. Then there are the spices: chilli, cumin, coriander and ginger, to name but a few. Finding a complementary white wine means looking for a relative heavyweight. Try out a South African Chenin Blanc, a Chilean Sauvignon Blanc or a white wine, such as a Chardonnay, from the Languedoc region of France. You need enough richness to add interesting flavour, and enough acidity to counteract what food scientists would call the 'mouth feel' of that oiliness on your palate. So avoid a 14% California Chardonnay, or it could get rather too intense.

If you fancy a red wine, one of the best with curry is red Rioja. Aged in tangy American oak, it has a warmth and pungent fruitiness that survive the onslaught of even fiery chilli pepper. Sample it with a lamb rogan josh or a chicken tikka masala (a dish said to be even more popular with the British than fish and chips). An alternative would be a smooth, faintly spicy Nero d'Avola from southern Italy – for example, Puglia or Sicily.

Thai and South-East Asian Cuisine

～

From peanuts to peppers, coconut to curry, there is a wealth of rich flavour to be found in Thailand, Indonesia, Malaysia, Vietnam and Cambodia. Some of the region's finest dishes are based on fish; with these, try a rosé from Provence, a pink Syrah from South America or a Loire Valley version from Anjou or Saumur; or try off-dry Riesling from New York State, Australia or Germany. Alsace is another great source of fruity white wine made with Riesling, Pinot Gris or Muscat. If you prefer contrast, try a mature Beaujolais from Juliénas or Morgon; alternatively, Gamay adds zip to this cuisine.

Chinese and Japanese Cuisine

～

These cuisines offer a huge repertoire, encompassing the extreme simplicity of sashimi, and the rich redolence of shark's stomach or sea urchin, as well as various contrasts of sweet, sour and hot. Try a good Cava to cover all bases, or be more specific with a Muscadet or Galician white wine for sushi and other simple fish (the tart acidity is a delight with seaweed); light red wines, such as Bordeaux Supérieur or basic red Burgundy, with fried dishes or dim sum; and late-harvest Riesling or Australian Semillon with the very hottest, richest food.

Middle Eastern Cuisine

∿

A classic meze in Lebanon, Greece, Turkey or North Africa includes a sultry,
spicy tang from sumac and pomegranate molasses. Cut through some of the
'barbecue' flavour with the clean, fresh acidity of such a wine as Gavi from
Piedmont in Italy, or Chablis from France. Alternatively, try a local red wine
from these regions; they are a great complement to the depth of smoky
flavour in such dishes as baba ganoush and tagine.

Complementing 'Impossible Ingredients'

Asparagus	This delectable and delicate vegetable has an affinity with the Viognier grape, and the two can combine very happily. Avoid serving asparagus with a very dry wine, or any red that has much tannin. A light, fruity red, such as Bardolino, is fine.
Artichoke	This is not dissimilar from asparagus in the way it neutralizes the complex flavour of fine wine. The globe artichoke is a huge favourite in France and Italy. Try it with any Sangiovese wine (even Chianti), or, for a more subtle approach, select a complementary, rich Chardonnay from Australia.
Vinaigrette	Purists of haute cuisine aver that any salad dressing made with vinegar should never be served with wine. But it is worth attempting: select a very crisp and tart wine, such as Vinho Verde, if you are serving a salad as a main course, or try some red wine that packs a good punch of fruit and acidity, such as Argentine Malbec – then stick with this red with the cheese course. In France, it is common to serve salad alongside the cheese.
Chocolate	This constitutes a mouth-coating treat that defeats most wine but is great with fortified, slightly sweet styles, such as Maury, from the South of France; alternatively, offer a sweet Moscato di Pantelleria or an equivalent Muscat wine from Australia or California.

Visiting the Vineyards of Chile and Argentina

In the above map of Chile and Argentina, the highlighting shows those countries' wine-production regions that are discussed in this feature.

Chile

This is a country that has just about everything a grapevine could desire: long, hot sunny days, cool nights and, thanks to a lack of vine disease, the very real possibility of producing organic wines without too much effort. For the visitor, it offers the bonus of dramatic natural beauty and the chance to taste wine and to ski or swim, all within a matter of hours. As you arrive in the capital city, Santiago, the peaks of the Andes surround you at once, and you need simply drive south down to the Central Valley to find fantastic Pacific beaches and then the vineyards, overlooked again by those stunning snow-capped mountains.

The early Spanish settlers here realized rapidly that this is a perfect region for the vine; today there are well over 110,000 hectares (271,815 acres) under cultivation. Although the main Central Valley is still the heartland, there are all sorts of experimental areas where vines are grown, such as the Elqui Valley and the Malleco Valley. From Santiago, the closest area for the wine tourist is the Casablanca Valley, where there are many long-established wineries, including Cousiño Macul. Its historic tasting cellars are right on the edge of the city, although the new facilities are farther afield.

Continue south towards Colchagua, which has been a wine-growing area for a very long time and has a great reputation for its red wines, notably those made with Cabernet Sauvignon and Merlot. This region is well set up to receive the wine tourist, and there is an established *Ruta del Vino* (wine road) with some fourteen wineries taking part. For a sophisticated and elegant look at the industry, stay at the Casa Lapostolle winery, where you can indulge in fine food and wine tasting. This is home to one of Chile's most sought-after wines: Clos Apalta, made with advice from famed French consultant Michel Rolland.

Other wineries well worth a visit include Viña Montes, where you can taste not only Bordeaux-style varietals, but also some fine Syrah and Carmenère – the exciting wine of the moment in Chile, which has huge potential. Another Carmenère specialist is Viña

Casa Silva (see Interview, p. 71), which offers not only wine tasting, but also the opportunity to watch polo and traditional Chilean rodeo horseriding skills in a specially constructed arena.

Another very well-organized wine-tasting facility is that belonging to Miguel Torres, of Catalan fame (see p. 50); he came to Chile as long ago as 1979. Now this winery makes terrific, drinkable pink wines that travel all over the world, as well as a full range of white and red wines. Compare and contrast these with Torres's Spanish wines in his conveniently located tasting room in Curicó.

Returning to Santiago, do not miss the impressive Concha y Toro winery, the largest in Chile, which has beautiful gardens and provides the opportunity to visit its vineyards as well as the winery. Nearby is its glamorous Almaviva winery, which is a joint venture involving first-growth wine from Château Mouton Rothschild of

LEFT: Casa Silva's Lolol vineyards specialize in Viognier grapes, a white variety not generally associated with Chile.

ABOVE: Vines grow right up to the terrace at this Colchagua winery.

Bordeaux. Drive a little farther north into the Maipo Valley to see more lovely gardens at the Undurraga winery in Melpilla. If you plan to taste Argentine wine as well as Chilean, then there is a spectacular drive through the Andes awaiting you, towards the Mendoza Valley. En route, enjoy the lovely setting and excellent wines at Viña Errazuriz in the Aconcagua Valley; the red wines made here challenge the French at their own game and have done so for many years (the winery was first established in 1870).

ABOVE, CLOCKWISE FROM TOP LEFT: Wines are judged and evaluated by managers and family members at Casa Silva, Chile.

Horses are frequently part of life in Chilean and Argentinean vineyards, and stud farms co-exist with the vines.

A Chilean 'winemaker in the making' sports traditional costume at an event celebrating both the horse and the vine at Casa Silva.

Some of the world's finest polo-players hail from Argentina and Chile; and some are members of wine families.

OPPOSITE, LEFT: At Argentina's Estancia de Cafayate, a shared ownership scheme means that people from different parts of the world are able to come on holiday, be involved with the estate, play golf and ride horses.

OPPOSITE, RIGHT: The white grape of Torrontes vines in the Calchaqui Valley near Salta in Argentina has a fragrant, distinctive style and does particularly well in this region.

Argentina

Settlers from Chile made the crossing into Argentina past towering Mount Aconcagua, the highest peak in the Americas, as long ago as the sixteenth century. The Argentinian wine industry has long been centred on the Mendoza Valley. Many of the vineyards are located high on the Andean slopes, in growing conditions that perfectly suit the Malbec grape, originally from Bordeaux but now probably best known in its Argentine incarnation. As in Bordeaux, Malbec lends the wines depth of flavour and a long-lived austerity. The red wines of both Chile and Argentina are frequently singled out by scientists for their health-boosting polyphenol content (see 'May', pp. 73–75), a result of the intense sun that strikes these mountain vineyards. This intensity superbly complements the typical barbecued steak that is so beloved in both countries, and served on a grand scale.

Wineries around Mendoza that merit a visit include Bodega Catena Zapata, which resembles a Mayan pyramid and offers some great, stylish wines to taste. The Familia Zuccardi winery is also quite an enterprise, with a strong focus on attracting the wine tourist with classic-car tours, olive-oil tastings and plenty of slightly eccentric experimental wines, including Viognier, Zinfandel and Gamay. If you venture away from these more outward-looking locations, a sound knowledge of the local language and good map-reading skills may be needed, as wine tourism is a new concept for many of the older generation of Argentinian winemakers.

Fairs and Festivals

April generally sees the end of the Lent fasting and the celebration of Easter in Christian communities all over the world, and wine is usually a part of these festivities.

Chileans hold *fiestas de vendimia* (grape harvest festivals) in all the country's wine-growing areas, and the largest is in the city of Curicó: curicochile.com

In France, the city of Mâcon, just to the south of the Côte d'Or in the Burgundy region, is the site of a competition for wines from all over France – the Concours des Grands Vins de France – usually in late April but sometimes in early May, accompanied by a wine fair: concours-des-vins.com

In Italy, the giant Vinitaly wine fair in Verona is an April fixture, showcasing the bewildering variety of Italian wine and its colourful producers, as well as hosting tastings of all sorts of delicious Italian food: vinitaly.com

BELOW: At the Museo del Vino in Maipu, Argentina, visitors learn about the integral role horses and oxen played and continue to play in the wine industry here.

BOTTOM: Barrels for ageing red wines in Argentina's Mendoza province; 80 per cent of the country's wines are produced in the region, mainly from Malbec, Cabernet Sauvignon and Syrah grapes.

Mario Pablo Silva

Managing Director of Viña Casa Silva, Chile

The Viña Casa Silva winery, in Chile's Colchagua Valley, is the oldest in the region, and some of its wines have received international acclaim.

casasilva.cl

Casa Silva first became known on the export market in 1997. Was this the beginning of its history?

Actually, the story begins as long ago as 1892, when my ancestors arrived in Chile from St-Émilion in the Bordeaux region. They brought vine cuttings with them. Over the years, the vineyards thrived, but the wines were sold in bulk, and it was a relatively closed economy in Chile; there were few export opportunities until we established Viña Casa Silva in 1997. I joined the company in 1998. Prior to that I worked at a senior level in a major paper export company.

What would you say is your company's winemaking philosophy?

After all those years growing grapes, we have a deep respect for the concept of *terroir*, and we are sure that we can make some of the finest-quality wines in Chile. We are especially interested in working with the Carmenère grape, which shows huge potential in Chile [this grape came from the Bordeaux region, and is still cultivated there on a small scale]. At the moment, we are studying micro-*terroir* sites with a professor from the local University of Talca, and our winemaker has created some very interesting separate barrels of Carmenère from different corners of our finest vineyards. We hope to find a 'super clone' of this grape that will give great quality, because we really want to be the ambassadors for this variety. So much Cabernet Sauvignon and Merlot are grown in Chile, but we feel that Carmenère is our special wine. It also has a very high antioxidant content, so we are sure it is good for the heart as well!

Apart from winemaking, I hear that you are a champion polo-player and that Casa Silva in Colchagua offers various diversions for the wine tourist. Tell me more.

The family is very involved with our wine tourism and with the winery: we don't live in some remote mansion. If you visit Casa Silva, you will see us with our horses. There is even a rodeo stadium and a polo ground there, plus our boutique hotel (in our old family house), a restaurant and wine bar. This means the visitor can stay with us and get the chance really to understand our wines and our vineyards, while enjoying all the other activities.

What do you feel is the image of Chile as a wine-production country today?

I'd like to say that it is now seen as safe, secure and constantly improving. There is still a great deal of work to do to improve the image both of Chile and of its wines. The mid-priced wine range is where we should focus our efforts to interest the wine enthusiast. We are also looking at new, cooler climate regions for wine production, down towards Patagonia, where grapes work hard to survive, but may yield very good quality in the future. As a company, we want to work towards sustainability overall, and a very responsible attitude to our employees, our vineyards and our visitors!

May

Wine and Health

....

*...use a little wine for thy stomach's sake
and thine often infirmities*

St Paul, I Timothy 5:23

*I often wonder what the vintners buy
One half so precious as the stuff they sell*

The Rubaiyat of Omar Khayyam

Above: This stylish Art Nouveau poster by Orazi encapsulates the concept of wine as a hedonistic and life-enhancing beverage.

Opposite: Stunning, rampart-like terraces contain the Rheingau vineyards at Assmannshausen, Germany.

In the vineyard, May is a month of contrasts across the two hemispheres. In the southern hemisphere, as autumn sets in the leaves change colour, mellowing from green to gold and finally to red. By contrast, in the northern hemisphere, May is a time of renewal: fresh shoots are seen on the vine, and fear of frost damage is receding, to be replaced by anxiety about the future flowering. As the new growing season gets under way, it is intriguing to consider the role of wine in our personal well-being. From the apostles to the Romans, from the Greeks to the Persians, there seems to be have been a consensus throughout history that wine can have beneficial effects beyond the simple, hedonistic pleasure of drinking a glassful. The principal reason that Europe was liberally sprinkled with vines during the Roman occupation was not for social purposes, but so that wine could be used as a medicine: wounds were dressed with wine and honey, or wine was added to water that was in need of purification. These functions for wine are still valid, but there is much more to discover about the role of wine drinking in our lives and its effect on our health.

Experiments carried out in which a small amount of wine was offered to care-home residents each day show that the use of sleeping pills and other drugs fell almost at once. If taken in moderation, wine is calming to the mind, and it does seem to have the effect described

by St Paul, of helping us to digest rich food. There has also been research by experts, including Prof. Roger Corder of the William Harvey Research Institute in London, that indicates that wine drinkers often live longer than total abstainers, and that beer or spirits do not have the same beneficial effect. This suggests that it is not the alcohol that is doing good: it is something specific about wine.

The medical profession has had a lengthy relationship with the wine trade, and there are many instances of doctors owning vineyards and making wines. One of the most striking is in Australia, where the wine industry was virtually founded by two doctors who had trained in London before emigrating in the first half of the nineteenth century. Dr Henry Lindeman (1811–1881) was a great believer in the therapeutic value of wine, and he first grew grapes in the Hunter Valley of New South Wales. Dr Christopher Rawson Penfold (1811–1870), meanwhile, imported French vine cuttings and planted them near Adelaide. By the 1920s it was claimed that one in two bottles of Australian wine came from his winery (see the interview with Peter Gago of Penfolds winery in 'December', pp. 204–205), and his and Lindeman's names are still dominant in the trade today.

In 1979 a study published in the British *Lancet* journal showed that middle-aged men seemed to be healthier in regions where wine consumption was common; the highest death rates came in countries where beer and spirits were favoured. Then came a real bombshell, on the American television programme *60 Minutes*, in 1991: Dr Serge Renaud put forward his revolutionary theory, which came to be known as the French Paradox; Dr Renaud also published his findings in the *Lancet*. In a nutshell, he stated that regular wine drinking (especially of red wine) could account for the fact that the French had lower rates of coronary heart disease, despite a diet generally high in saturated fat. Studies were carried out across the regions of France, and, most curiously, revealed that the inhabitants of the Gers, in the heart of Gascony, seemed to live longest of all. This was odd, because the diet in the Gers regularly includes pâté de foie gras and many other dishes made using goose or duck fat. There was a storm of reaction to the results, and the subject remains very controversial, but other studies carried out in Crete and Sardinia on populations with dozens of centenarians have seemed to reinforce the value of red wine. The medical cliché is that we are 'only as old as our arteries', and wine may go some way to counter the negative influence of fats and rich ingredients in our diet.

There has been further research to determine what it is about red wine that makes this apparent miracle

occur. Prof. Roger Corder, in his book *The Wine Diet* (2006), makes a very good case for the value of polyphenols, substances found in all red wine. Among these, he singles out a group of flavonoids called procyanidins. There are masses of procyanidins in young red wine, and also in cranberries, apples, pomegranates and high-quality dark chocolate. If his belief is correct – and he has evidence to back it up – basing a diet on these delicious ingredients could help us to avoid stroke, cancer and heart disease in future life.

Other researchers are passionate about the virtues of resveratrol, which has been alleged to reduce inflammation in the body and thus to protect from disease. It is found in the skins of grapes, and is there to protect the bunches from attack by moulds and from the damage that can be caused by excess sunlight. It has been suggested that resveratrol is the reason why wine seems to have a protective effect on our bodies. Prof. Corder says that the amount of resveratrol in most red wines is not sufficient to confer these benefits, but there are plenty of researchers who disagree. It has recently been claimed that this polyphenol acts as an antioxidant (as they all do) and that it is powerful enough to counteract raging infections, such as septicaemia and peritonitis, which can set in when the appendix is inflamed.

All of which seems to lead us back to the Romans, who knew the value of red wine in treating war wounds; its natural antioxidants combined well with the antibacterial qualities of honey. Today, Manuka honey from New Zealand is still much prized for this purpose, and it may be that we should start thinking of wine more creatively. If wine really can delay the ageing process, by keeping our hearts healthier, then perhaps an infusion of grapeseeds is the best possible skin treatment. Only time will tell.

Meanwhile, the general advice appears to be that drinking one or two glasses of red wine per day, taken with a balanced diet of fresh foods, is likely to bring health benefits. It is, coincidentally, also very pleasurable and can be socially beneficial. In these days, when the wine trade is being battered by calls for restraint on labelling, higher taxes per bottle and warnings of the dangers of overindulgence, it is good to have some positive news for those hard-working wine producers.

Finally, the shattering of a myth: wine consumption is not the cause of gout. If port drinkers traditionally suffered with this painful complaint, caused by a build-up of uric acid in the joints, it was because port contains a shot of grape *spirit*. Beer, spirits and red meat are to blame, it seems, so the wine drinker can rest easy.

Wine Tasting at Home

Organizing an authentic wine-tasting experience at home is an excellent excuse for a party afterwards – but a world away from the days of the wine-and-cheese party, with its awkward connotations of cocktail sausages, cheese on sticks and stilted conversation. Now, you can start the evening with a dash of competitiveness and culture, and then enjoy what remains of the wine, ideally with some complementary food.

Tasting wine is easier than it seems; what you are seeking are a few key characteristics that may stay in your memory and help to build a 'library' of sensory experience associated with certain wines. This is entirely personal, and it is very important to let your imagination run free. Start with the conventional phraseology of the wine taster, then improvise a few riffs of your own. If possible, write down a phrase or two and keep those notes. Soon your nose will start to pick up cues from a single sniff at a glass. This skill is highly useful in expensive restaurants, as you pass your nose thoughtfully across the tasting glass, then set it down. As your taste buds simply confirm what the nose has sensed, actual tasting is not essential unless you suspect a fault. This is a classy way to show a sommelier that you are clued up about wine.

But to get back to your party: it is worthwhile planning a theme, rather than simply encouraging all your guests to bring a bottle. Think of wine by style (see 'June', pp. 89–91) and offer a selection of light, dry whites or rich reds; or select a particular vintage that has had good reviews, and ensure that all the wines carry the same date. In this age of varietal wine (see 'February', pp. 29–32), it is always interesting to compare wines produced from the same grape but from different continents. So a Cabernet Sauvignon from California can be set against a French St-Émilion or a red wine from the Veneto or Valencia.

On a more practical note, price may be a good starting point. Challenge your group to find wines that are all under a certain cost per bottle, yet that fit certain criteria. For instance, finding a good sparkling wine for a special occasion can be a very expensive undertaking if done alone. In a group, however, it is perfectly possible to compare a dozen wines in an evening, and everyone in the room benefits: the warm glow that ensues is not just those bubbles tickling the stomach lining.

Talking of the stomach, never eat just before or during a tasting. The only exception would be to provide a very plain, unsalted cracker to nibble if required. Provide some water too, as the palate benefits from a refreshing sip between tastes. This should be still water; many fizzy mineral waters pack quite a salty punch, which affects the taste of wine. Professional wine tasters also avoid smoking, wearing scent or cologne – and lipstick. These aromas can alter the 'nose' of a wine quite radically.

So, the group is gathered in a state of high seriousness. To preserve this atmosphere, provide spittoons: either empty bottles with plastic funnels, or a couple of empty ice buckets. Each taster will need a single glass for whites, and another for reds. The best shape of glass for tasting is a 'tulip', which offers scope

Home wine tasting requires little other than appropriate glasses: the 'tulip' shape, with a bulbous body and narrower head, will concentrate the wine's aroma.

to swirl and release those aromas and then focuses them at the top of the glass. If any wine seems especially persistent in the glass after tasting, just sluice it round with some of the water on the table and empty this into the spittoon.

Cover the table with white paper and write directly on it to indicate a place and number for each wine. As a general rule, work from left to right, and therefore make sure that any lighter styles of wine precede known alcoholic 'monsters'. If you fancy a 'blind' tasting, slip each bottle into a numbered paper sack and tie it shut: ripping off the paper to reveal the wine is surprisingly satisfying after the tasting game is over.

Each bottle should provide about fifteen tasting samples, so pour only a large splash of wine in the glass. Hold the glass at a slight angle and look at the colour of the wine against the white table cover (or your spotless white cuff?). Ensure that there is enough light to determine if your Chablis is a delicate straw colour (as it should be) or a rather sinister oily yellow (which would probably indicate unwelcome oxidation in this style of wine). Think jewel colours – garnet or ruby tones – for red wines. For whites, begin with pale straw and advance to shades of gold for sweeter wines.

It is unlikely that the wine will be anything other than clear and bright; modern winemaking is so technically accomplished that mystery deposits in the glass are rare. Some organic wines or very elderly red wines may throw some sediment, which is generally quite harmless. Hold the glass by the stem or the base (to avoid warming the wine through the heat of your hand) and swirl it gently, then apply your nose. Some tasters close their eyes and allow the aromas to 'surprise' them. Note down a couple of words to summarize what you find: fruity, perfumed, green, woody, fragrant. Then take a sip, swirl the wine up and down the sides of your tongue, detecting the acidity as it goes. The tannins will be apparent at the front of your mouth, just like those in brewed tea.

After spitting out your sample, add more thoughts to those simple words already noted. Does the wine make you think of your dog returning damp from the woods? Or does it have the scent of a summer meadow? How about the interior of an antiques shop, with its beeswax and aged wood? Or even a beach at sunset, with a salty whiff of the sea? Let your imagination go, and enjoy the experience. It can be shared, or simply privately relished. Who said education had to be hard work?

Visiting the Vineyards of Germany

1 Mosel Valley

2 Rheingau/Rheinhessen

3 Baden-Württemberg

The map above shows the wine areas of Germany that are discussed in this feature (see key, above). The numbering follows the order in which the areas are covered.

A great deal of misunderstanding surrounds German wines. The general perception is that most of them are sweet and lacking in any specific character; or that they are not suitable to match with food. To knock this theory on the head, go to Germany yourself, if you can, and see how completely integrated into the culture fine wines are. It is quite a large country, with a variety of regions to choose from, of which the Mosel and the Rhine valleys are the best known. One thing is for certain: wherever you travel, the standard of welcome and wine tourism is very high, with great places to stay, and opportunities to taste.

Germany's oldest city, built originally by the Romans, is Trier, on the shores of the Mosel river. It is a UNESCO World Heritage Site and the perfect place to begin a visit to this scenic region. Drive out along the meandering roads towards Bernkastel and you will be astonished by the way vines grow on the vertiginous slopes above, only accessible on foot: there is no mechanical harvesting here. Visit Bernkastel's helpful Weinmuseum, and its Vinothek, where you can taste local wines. For the very finest from the area, seek out bottles from Egon Müller's Scharzhofberg (in the subregion known as the Saar) and Maximin Grünhaus's Abstberg from the Ruwer (another tributary of the Mosel). The

S.A. Pruem winery in Wehlen is a great place to spend some time, as it produces excellent Riesling and also has an attractive guesthouse. Many of its wines are made in a drier style, and it is interesting to match them with local dishes, such as fish from the nearby rivers.

The Rheingau is another beautiful, historic and atmospheric region, full of classic wineries and glorious scenery. The village of Hochheim is reputed to be one of the earliest sites where Riesling was planted, and British wine lists of past centuries always featured a selection of 'hock'. Top among local attractions has to be the spectacular Schloss Johannisberg at Geisenheim; this is a modern reproduction of the old castle (destroyed in the Second World War), but the cellars beneath are the originals, full of giant oak casks and lit by candles. Another fine estate with excellent Riesling is Schloss Vollrads at Oestrich-Winkel. One of the most enjoyable ways to spend time here is to take a river cruise, either

ABOVE: The uneven terrain at Wehlener Sonnenuhr in the Mosel region makes it is easy to appreciate why grapes here need to be handpicked, meaning that labour costs are high.

BELOW: The Mosel river forms a wide loop near Trittenheim. The classic grape of this region is the Riesling.

OVERLEAF, LEFT: The historic Rheingau estate of Schloss Vollrads, where wine has been made for over 800 years, has developed a reputation for fine wine and is noted for its Eiswein.

OVERLEAF, RIGHT: The medieval atmosphere is echoed in the ancient walls surrounding these Rheingau vineyards.

BELOW AND BOTTOM LEFT:
The cellarmaster at Prinz von
Hessen's wine estate selects some
bottles from a mature vintage for
tasting. The winery's Dachsfilet
is produced from Riesling grapes.

BOTTOM RIGHT: A saying goes,
'Let the vine see the water', and
these vines have a glorious view
of the Rhine and the perfect
aspect for gaining maximum
exposure to the sun.

for just one day or for an entire week, to appreciate
the way the vineyards are integrated into this striking
landscape, loved by Goethe and many other writers.

In the Rheinhessen region there is substantial
production of sparkling wine, known as Sekt, and it is fun
to take a tour of one of the cellars. This is the home of the
notorious Liebfraumilch, and it is only in recent years that
local wines have started to improve, thanks to the efforts
of some dynamic young winemakers, such as Keller and
Kühling-Gillot. The town of Nierstein is renowned for
wine: look out for producers in the vicinity, where you
may get opportunities to taste the wine and make up
your own mind. There are also interesting wines to be
sampled in the Pfalz, in the shadow of the Haardt
Mountains, not far from the French border. This is a
relatively dry and warm region, where the Riesling has
an easier time ripening than in the extreme conditions

of the Mosel-Saar-Ruwer. Visit Deidesheim and the estate of Reichsrat von Buhl to see how these wines can taste so different from one aother. In some ways, they resemble the fine white wines of Alsace, across the Rhine in France, possessing plenty of ripe, fruity character, yet with a dry finish.

Away from the river, and into the Black Forest to the south, are the vineyards of Baden-Württemberg. These are not Germany's finest wines, but this is a glorious region for visitors, with opportunities to hike, eat and drink to your heart's content. A great many wines are made at the giant local co-operative, the Badischer Winzerkeller, and there are some very drinkable red wines to try, made with Pinot Noir (called Spätburgunder in Germany). If you prefer white, try a glass of rich Pinot Gris (Grauburgunder) with one of the region's calorific but irresistible fruit tarts.

BELOW: A lone watchtower stands guard amid densely planted vines in the Rheinhessen region.

RIGHT, TOP AND CENTRE: The Prinz von Hessen wine estate in the Rheingau runs the Schlosshotel Kronberg; its elegant dining room is a fitting setting for its wines.

ABOVE: Spring growth on these vines in the Nahe region, to the east of the Mosel Valley, could be threatened by hail or late frost.

Dr Clemens Kiefer

Wine Director of the Prinz von Hessen wine estate in the Rheingau, Germany

The Managing Director of the company is Donatus, Prinz von Hessen, whose family can trace its lineage back to the thirteenth century. Wines produced by the Prinz von Hessen estate have won many international awards.

prinz-von-hessen.de

Tell me something about yourself and your winemaking philosophy.
I was born in the Rheingau, near Johannisberg. I've been interested in winemaking since childhood. My uncle had an estate and my father was Professor of Viticulture at Geisenheim University. I served an apprenticeship as a winemaker, then I studied business administration. In October 2005 I started work with Prinz von Hessen.

For me, a Riesling should be refreshing, lively and fruity, with a nice and long aftertaste. As we make it at Prinz von Hessen, I am convinced that you can show it off at three quality levels.

Our first one is the 'H Riesling'. This is a Riesling even for consumers who do not know too much about the style. The next level is the Prinz von Hessen Riesling Kabinett dry. It's a wine with a beautiful mouth feeling, a nice aftertaste and a great mineral aroma; an aristocratic wine. Our top Riesling is our Prinz von Hessen Riesling 'Dachsfilet'. This wine is powerful but not exhausting; it has various aromas. The taste combines gentle spice and matured peach notes within a tight structure.

As a niche, we have our noble sweet wines, such as Johannisberger Klaus Riesling Beerenauslese. This is a honey-coloured wine that fascinates the drinker with its fruit. Exotic fruit flavours come together to the nose; they are supported by a caramel honey note on the roof of the mouth – all of which harmonize wonderfully.

Do you share the opinion of some wine critics that Riesling is the noblest of all white grapes?
Yes I do. That is the reason why I work at Prinz von Hessen. The Riesling is the only variety with which you can produce the different qualities and sugar levels I have just explained.

How do you see the future for German wine in the UK and US markets?
I see a great future for high-quality dry Rieslings. When these wines are made, the consumer can have fun with them!

Finally, would you mind choosing just one wine from your portfolio and telling me why it is your favourite?
It really depends on the occasion. Our Prinz von Hessen Riesling Kabinett dry is such a refreshing wine, I like to drink it when I am thirsty. Our Dachsfilet is my favourite for the colder season, when it is wet and foggy here in Rheingau.

The louche streets of Naples can hide its cultural heart; this is a complex and intriguing city to visit, and the Vitigno wine fair provides the perfect excuse to discover its charms.

In southern France, join in the fun of Apt's Grande Cavalcade of wines: ot-apt.fr

Enjoy some late-spring sunshine in Naples, Italy, where the Vitigno wine fair is in full swing, celebrating wines of the Mediterranean: vitignoitalia.it

In the UK, the London International Wine Fair (LIWF) is a major international wine trade fair: londonwinefair.com

June

Wine by Style

· ·

🍷 In the southern hemisphere, the vines become dormant by June, but, by contrast, in the northern hemisphere June is a wonderful month to visit a vineyard; the weather is warm, the leaves on the vines are a glorious apple-green, and the fragrance of *la vigne en fleur* waxes seductive. Alternatively, in a year of unpredictable weather, there may be strong winds rattling the posts and wires, heavy rain causing rampant vine growth, and sudden infestations of unwelcome pests. That is the life of a winegrower, and as a result it is probably a mistake to categorize wines from any region in a fixed fashion.

To say that all Bordeaux claret is tart, tannic and slow to develop in bottle would be a nonsense if confronted by, say, an example from the scorching summer of 2003, when these wines hit record alcohol levels and, when tasted, can be as warm and approachable as any of their California cousins. Similarly, to classify all wines made with Muscat as sweetish, scented and soft would be to ignore the amazing dry white wines made with this variety in the marginal climate of Alsace.

LEFT AND ABOVE: The brilliant green of early vine growth in June contrasts with the colourful wild flowers that show that this vineyard's owner avoids the use of herbicide.

OPPOSITE: This typical Bordeaux above-ground cellar, or *chai*, houses traditional glass-lined cement fermentation tanks and oak barrels for storing the wine.

Oak plays a key role in determining the style and ability to age of any wine, red or white.

Given the complexity of the subject, one might wonder how to start organizing wine by style. Many restaurant wine lists now attempt to offer diners a choice of 'Rich Reds' or 'Fruity Whites', and this is a good place to begin when confronted with unfamiliar wines. There are some wines that seem to lend themselves to this type of classification. Chablis is generally light, dry and crisp; Australian Shiraz is ripe, robust and full of flavour. Yet there are also plenty of exceptions to these 'rules'. Grand Cru Chablis that has been oak-aged and stored for seven years or so will have all the delicate, rounded charm of a white Burgundy; sparkling red Shiraz is almost flowery in the mouth.

To get to the heart of how to sort out wine by style, cast aside the cliché and think about the absolute basics. In essence, we are seeking a wine that will pair happily with whatever we intend to eat. Most wines taste better with food, and vice versa. Just one glass with that panini at lunchtime raises the meal to the status of enjoyment, rather than being simply fuel. Likewise, unsurprisingly, wines from particular regions seem designed to accompany local dishes, which is why many Greek wines somehow taste better on the island rather than at one's own dinner table, and the wines we sample after skiing do not quite hit the spot back home. Appreciation of wine is significantly affected by ambience, which is hard to replicate.

Instead, start by thinking about the five elements of any wine: acidity, sweetness, tannin, oak and alcohol. These form the structure of the wine, and if you think about them on a sliding scale, then a well-made wine would hold them all in balance.

Acidity

Any good wine has a fair measure of acidity, which gives the wine its charm and finish, and counterbalances any tendency to sweetness. Most of the world's great wines are notably acid on the palate; this tartness adds 'zing' and is found even in fine dessert wines, such as Sauternes or German Beerenauslese. It is easier to produce a wine with good acidity in a cooler climate, and white wines display it more prominently than reds. This is why Gavi di Gavi from northern Piedmont in Italy is famed for its citric tang, and the white Grillo of Sicily tends towards a ripe, oily style. In red wines, acidity is transformed when it meets up with a well-chosen dish; many drinkers describe red wine as 'sour', but very often this is because they have sampled the wine without food. The wine trade is said to 'buy on apples, sell on cheese', because if a wine can still taste good after taking a bite from a Granny Smith, it really has character. On the other hand, almost any red wine, however young and tart, will taste smoother when it encounters the lactic acids in cheese.

Sweetness

The sugars in a bottle of wine are described as 'residual', because they stay there after fermentation. This need not mean that the wine strikes you as being of a sweet dessert style. If the wine is acid enough and has good structure, then the sugars just melt into the whole experience of savouring the wine in one's mouth. Sweeter wines may be called 'late harvest', and far more natural sugar is found in wines produced in a reliably warm climate, such as that of California, Chile or South Africa.

Tannin

Natural tannin comes from the skin and seeds of the grape. It adds a hint of bitterness to many red wines, which is ideal when you are serving meat, and a disaster if you are offering smoked salmon – hence the general rule about serving white wine with fish. Tannin is particularly noticeable in younger wines, as the ageing process softens the tannins. Unless you like mouth-stinging astringency, look out for wines that are a few years old – or choose red wines from benevolent climates, such as Chile or Sicily.

Oak

Many fine wines are oak-aged, and too many ordinary wines are 'oaked' with the addition of oak chips or even added oak extract in an attempt to give them a veneer of class. The vogue for heavily oaked Chardonnay seems to be passing. A well-kept wine with some years of age in bottle will show some oak character, but it blends in with the acidity, tannin and sweetness to create that wonderful velvety quality, known as bouquet, in the mouth. At this stage, oak-barrel ageing pays off. Prior to that, oaked wines can be attractive on the nose, with their vanilla or coconut aromas, but may be tough on the palate.

Alcohol

Pure alcohol is absurdly sweet; no wonder it can be addictive. The oily 'legs' we see on the side of a wine glass are some indicator of how much alcohol is present, and many wines are now significantly higher in alcohol content than those that went by the same name in previous decades. Thanks to the influence of New World wines, consumer taste has veered to this style of winemaking. That viscous alcohol/glycerol coats our palate, so it needs something substantial to balance that sensation; for example, a meal with full flavours rather than just a packet of something salty.

Taken all together, these are the principal elements of wine style. Study the information on your next bottle: check the alcohol level, and see if the label on the back of the bottle mentions anything about late harvest, if the vintage is indicated, and anything about how this ties in with the use of oak. Talk to your wine merchant, and ask to see samples of wines that might illustrate the contrasts: a well-aged French red wine versus a fiery, youthful Portuguese red from the Douro, for example, or a light, dry Muscadet versus a late-harvest Chardonnay. Restaurant menus sometimes can be a mine of information on this topic. Alternatively, consult the websites of top-quality firms, such as Berry Bros & Rudd (bbr.com) or Bibendum (bibendum-wine.co.uk). Even supermarkets are starting to see the commercial advantage of recommending wine by style. There is so much to learn about wine style, and, fortunately, it can be the study of a lifetime.

Wine and the Great Outdoors

....

Bread, wine and fresh air —
life's perennial pleasures.

Some day, there will be a scientific study proving that food and wine taste better when sampled outdoors; but whether this is true or imagined, the sensual pleasure of eating with our fingers and sipping from a glass of wine while out in the open air is an experience most of us find uniquely relaxing and enjoyable. The simplest version of this activity is the good old-fashioned picnic, improvised from the best ingredients to hand and accompanied by a suitable bottle of wine from nearby vineyards. This is the sort of holiday memory many of us treasure, and the reason we crave visits to real wineries on our travels. Worldwide, vineyard proprietors are realizing the attraction of dining amid the vines, and are offering organized picnics for their visitors, alongside more conventional winery tours and tastings.

On a larger scale, outdoor eating may consist of a complete lunch or dinner that is served to guests at a sporting occasion or a group activity, such as a motor rally or a friendly gathering by the shores of a lake, watching water sports. It is much more enjoyable to eat on the spot rather than having to go in search of food, spoiling the spontaneity of the moment.

Assuming your budget does not run to a hamper with some iced Champagne served by flunkeys, it is still possible to ensure a really great outdoor eating experience. Certainly a bottle or two of fizz does make a great impression. There is such an excellent selection of sparkling wine at affordable prices today that it could be fun to choose a wine to accompany the style of food: English fizz with pork pies and sandwiches; Cava with tapas; Prosecco with a selection from the nearest Italian deli; or sparkling Saumur with smoked trout and charcuterie. California fizz would lend itself to sourdough bread and cracked-crab salad; Australian sparkling wines make a great, fruity accompaniment to spicy noodle dishes from the Pacific Rim; and the tangy, sparkling white wines of New Zealand are delicious with fresh seafood, such as mussels.

Once the fizz is served as an ice-breaker, what next? Watch out for alcohol levels in your choice of wine. At lunch, a deliciously light wine is more quaffable and avoids the risk of all your guests falling asleep before any post-prandial play commences. German Riesling used to be a classic choice, and there are now so many good

examples of this grape produced elsewhere, from Australia to New York State, that this is an opportunity to give it a try. Another aromatic variety that lends itself to being made in a dry style is Muscat; look out for wines of this type from Alsace. If your picnic is spicy or perhaps rich in pâté and cold meats, then consider a fresh, fragrant Gewürztraminer, again from Alsace, Germany or the United States. Last (but not least) in this family of white wines is Grüner Veltliner from Austria, a truly refreshing and rewardingly clean-tasting style of wine that adapts happily to all cuisines.

For those who want to conjure up the spirit of a rustic French feast or an alfresco Italian meal, a good choice is a red wine that has just the right spirit of fun but not too much weight in the mouth; a Beaujolais-Villages, with its raspberry zing, or a Valpolicella from the Veneto. Cabernet Franc is a good variety to watch for,

and this shows at its best in the Loire Valley (Chinon, Bourgueil and Saumur-Champigny), and also impresses in Chile and the eastern United States. If elegance is needed in a red wine, then a fine Pinot Noir from Burgundy, Oregon or New Zealand is soft on the palate yet lingering on the taste buds.

Here are a few ideas for infallible picnic fare and wines to accompany it. Remember to choose screwcap bottles if possible, in order to avoid struggling to find the corkscrew (or forgetting to pack it back up), and make sure the whites are really well chilled in advance. To keep it all as eco-friendly as possible, use bamboo plates and cutlery that will compost down afterwards, and consider using glass tumblers rather than plastic for the wine; glass is also far better if you really want to relish the full flavour of the wine.

British Nibbles

Serve with Champagne, Mâcon Blanc and Pinot Noir

Quails' eggs boiled in the shell (hardy folk will eat the shells as well).
Tiny drop scones or blinis made fresh on the griddle that day, served with
caviar or lumpfish roe.
Smoked Scottish salmon rolled round cream cheese and chives.
Cucumber segments, part-peeled and stuffed with salty brown shrimp.
Mini pork pies and game pie.
A selection of sandwiches, including egg and cress, Cheddar cheese and English ham.
Warm new potatoes with melted butter and chopped mint, stored in a vacuum flask.
Strawberries, Scottish shortbread and Cornish clotted cream.

A Continental Affair

Serve with sparkling wine, rosé from Provence and a light red wine

～

Miniature quiche Lorraines, pizza and pâté with toast.

A selection of quality olives.

Ripe tomatoes stuffed with mild goat's cheese, sprinkled with chervil.

Parma or Serrano ham.

Plenty of crusty bread and good olive oil for dipping.

Hummus, taramasalata and tzatziki, with dipping vegetables.

Green salad dressed with lemon juice and oil, decorated with fresh flowers, such as borage.

Individual crème caramels, panna cottas and fruit compotes to finish.

A flask of excellent black coffee and some sugar stirrers.

Cheese and Wine

Serve with Chablis and Cabernet Franc

～

For the simplest of all picnics, just raid the cheese
counter for a selection of local fresh examples, add good
bread, butter, chutney or pickle and some tiny cherry
tomatoes. Conclude with fresh figs, sliced and dipped in
honey, and a single glass of Moscato per person.

Visiting the Vineyards of South Africa

OPPOSITE, CLOCKWISE FROM TOP LEFT: Vines line the Route 62 wine route by the small town of Ashton, by the Breede river.

In the stunning Franschhoek Valley, French Huguenot refugees first made wine in the seventeenth century.

Kanonkop, in the Stellenbosch region, is famed for its Pinotage red wines.

Microclimates abound in the Stellenbosch region with its rugged outcrops, giving a huge diversity of wine styles.

Wine producer Ken Forrester selected this tranquil Stellenbosch site for his estate.

These old winery buildings at the La Motte estate in Franschhoek are constructed in typical Cape Dutch style.

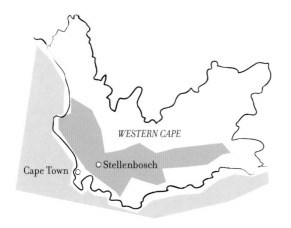

The map above shows South Africa's Western Cape province; highlighted is the province's main wine-production area, where most of the country's wines are made.

South Africa is a country that has undergone the most dramatic changes in recent decades and is now emerging as a sophisticated and well-structured wine tourism destination. The emergence of the 'Rainbow Nation' has brought great critical attention to the wines and a corresponding improvement in overall quality. Of course, the attractions of this part of Africa are not limited to wine appreciation, and the winery owners of the Western Cape are fully aware of how they can appeal to those who want to add to their experience of a great sporting event or a wildlife safari with a few leisurely days of wine tasting.

South Africa has some six hundred wineries and a huge wine output, which currently puts it in eighth place in the list of the world's wine-producing countries. Naturally, not all of this wine is of top quality, and there have been political and other problems besetting the country's attempts to build its image since the birth of

the 'Rainbow Nation' in 1994, when Nelson Mandela came to power. Nevertheless, tourism is hugely important to the national economy, and wine tourism is a great employer of local people. About 500,000 people work in the industry, and there are some very positive developments in the direction of encouraging black workers to take responsibility through the Black Economic Empowerment (BEE) movement, and through the official Fairtrade labelling of enlightened wineries. Ethical concerns are top of the agenda here, after all the years of apartheid and the resulting poor image that South Africa had in the eyes of tourists.

Now that South Africa is one of the top long-haul destinations for visitors from the UK, and that tourists from other countries are following suit, the wineries are starting to relax a little and develop further their already excellent range of facilities for the visitor. If you are planning to travel down to the Cape, the month of June is highly recommended by such locals as Ken Forrester (see Interview, pp. 100–101). The weather is usually excellent, and ideal for touring these spectacular vineyards.

The majority of the wines here are made within a couple of hours' drive of Cape Town; and the journey is gloriously scenic, featuring historic Cape Dutch houses set amid hills and valleys, and a range of welcoming wineries that also include excellent visitor facilities. One of the most famous is Groot Constantia just south of Cape Town, which has a lengthy history, dating back to 1685. It is curious to reflect that many of the seventeenth-century wineries in this region actually predate the famed châteaux of the Médoc in Bordeaux. The Dutch were the first to plant vines in Constantia, which has an ideal climate for viticulture; and the Boer winemakers are still prominent in the industry today. For quite some time winemaking was a state-controlled affair, and the wine quality suffered as a result, with too many white-wine vineyards yielding unexciting wine. In the past ten years, there has been extensive replanting (as much as 40 per cent

Opposite: The spectacularly sited family homestead at the Verdun estate in Stellenbosch was adapted to the needs of wine tourism and is now a tasting room.

Clockwise from right:
Widely spaced rows make access easier for ploughing, trimming and picking.

The Vergelegen estate in Stellenbosch is highly praised for fine Sauvignon Blanc, blended with Semillon and fermented in barrel.

Wildlife in all its variety is very much a part of the South African wine-touring experience.

of the entire vineyards of South Africa), and there is now more emphasis on quality and on fine red wines, such as the Pinotage; this variety, unique to South Africa, is pungently flavoured and dense in texture.

Stellenbosch, a pretty little town that is reminiscent of small towns in Alsace or Germany, is another key visitor centre. This is a prime vineyard area, and many of the state-controlled vineyards of KWV, a huge co-operative organization, were originally found here. As in Constantia, there are many opportunities to visit and to taste wine, for example at such wineries as Saxenburg. Wine and food matching is a key theme for South African wineries, and menus are beginning to reveal the influence of other countries, evolving beyond the simple basics of meat-based Dutch and British cuisine. But if you want a true taste of the Cape, try the typical biltong, a salted, spicy dried meat made from beef, ostrich or such antelope as impala; here is a match for sturdy red Pinotage wine. Another classic dish is

boerewors, a pungent sausage grilled on the barbecue (known as a *braai*). There is excellent fish, too, as the coast is not far away. In Cape Town itself, try the seafood or some spicy Cape Malay fare, which matches well with wines made with the typical local Chenin Blanc (also known in South Africa as Steen) grapes. These were once the most widely grown variety in the country.

Cellar-door restaurants offering diverse dishes are a feature of wine tourism here, and some wineries have their own goats, enabling them to produce excellent cheese and other ingredients that go straight to table and make for entertaining possibilities in combining wines and food. Eating out in this way is excellent value for money; along with the wildlife, the whales along the coast in Hermanus Bay, and top sporting events, such as world-class rugby and football, the classy food and wine is another reason why tourists are flocking here in their thousands.

Ken Forrester

a charismatic producer of South African wine

Ken Forrester's early career was in the hotel industry, followed by the restaurant business. He began work on his vineyard in Stellenbosch in 1993 and has since gone on to produce top-quality wines that have received international acclaim.

kenforresterwines.com

Tell me some history – how you came to restore your historic estate.

When I first came across the site in the early 1990s, what struck me hardest was the derelict state and the thought that 'someone' actually owned this piece of history that was more than three hundred years old – someone with apparently no respect for or understanding of the past. How could anyone possibly ignore the lives that had passed through here over the last three hundred-odd years? So renovation, or rather restoration, was very necessary. At the same time, obviously the vineyards, too, were a priority. However, unlike the homestead, they produced an income, so they were in reasonably good shape; very over-cropped, but after a year or two of severe pruning, that could be rectified. The first task was to find a sympathetic architect, one who would be happy to read the lines of history as well as accept that the house had to become a modern home while preserving the legacy of the past.

How did you select the grape varieties?

South Africa, specifically this part of the Cape along the coast east and west of Cape Town, has a predominantly 'Mediterranean' climate, and one would therefore say that surely this predicates exactly which grape varieties should do very well here.

We are located just 4 miles (6.5 km) from the coast of the cool Atlantic Ocean, along the False Bay coastline, at the foot of a magnificent mountain range that reaches 915 metres (3000 ft) at this point and summits in the mighty Drakensberg range of Kwa-Zulu Natal at about 3500 metres (11,500 ft; from where it runs from this southernmost tip to the very edge of the great African Rift Valley and the source of the biblical Nile river). This exciting landscape gives birth to a myriad of opportunities. And while Sauvignon Blanc and Chenin Blanc may be the best-suited whites for now, Cabernet Sauvignon, Syrah and even such Mediterranean grapes as Mourvèdre and Grenache are finding favour, and more and more wines are paying homage to this extraordinary climatic heritage.

I know your wines are widely exported. How do you feel the wines of South Africa fare in the UK and USA these days?

Generally, in these markets our South African wines, hitherto not necessarily portrayed as the pinnacle of quality – owing mainly to some sharp buying practices of the supermarkets, and also the desperate rush or need to 'normalize' trade after the isolation years – now represent extraordinary value. Hopefully, in the years ahead, these top wines will garner better prices and translate into better returns for the producers, and, in turn, better lives for all involved in the industry.

How would you describe the attractions of your region and your winery to the potential wine tourist?

The resounding opinion of every single person to whom I speak about the visual beauty of our wine regions is simply that 'there is no more beautiful wine region on Earth'! We have ocean, mountains, beautiful cities, an international airport on our doorstep, awesome shopping, breathtaking open countryside, rolling wheatfields, fantastic access roads and infrastructure, wonderful hotels and guesthouses, and some of the very finest wines and cuisine in the world at incredible value – what exactly were you looking for?

What plans do you have for the estate in the future?
We need to replace some of our older vineyards, and an ongoing replanting programme is in place, taking care to preserve and maximize the impact of the very special fruit from the oldest vineyards. We have recently started a very extensive composting process, as well as a stableyard with a dozen horses. A firm commitment to organic, sustainable viticulture; a small preserved wetland, newly fenced to provide a safe breeding-ground for rare wild antelope, which will ultimately be relocated to provide stock for game reserves around the country; and a small, intimate cellar-door operation with incredible mountain views, all set off by some memorable wines: these other plans are under way. We truly relish living what is a dream come true.

Finally, could you nominate just one wine from your range as a favourite, and tell me why?
I would have great difficulty singling out just one wine; there would have to be two. The FMC [an outstanding Chenin Blanc] was a true groundbreaker; it has helped to rewrite the potential of Chenin Blanc in South Africa and has won a huge collection of trophies, awards and very loyal followers. The Gypsy – a Rhône-styled blend of Syrah and Grenache – was also in its own way a groundbreaker, and was probably one of the very first blends of its sort in the country.

Below and bottom: Visitors throng the Bordeaux exhibition centre for the Bordeaux Fête le Vin festival.

FAIRS AND FESTIVALS

· ·

In Germany, June is the month for the 'Spargelfest' in the Rhine region, at which fresh, seasonal asparagus is paired with local white wine: germany-tourism.co.uk

In France, you can party at the Bordeaux Fête le Vin festival, celebrating the city's revival as a wonderful tourist destination as well as the home of many of the world's most famous wines. And in alternate years, there is also a big trade event in the city, Vinexpo: bordeaux-fete-le-vin.com; vinexpo.com

Milan is the site of the annual Mi Wine fair, a unique opportunity to try wines from all over Italy; details from the Milan tourist office: visitamilano.it

And in the Rioja region of Spain, there is a lively Battle of Wine (*batalla de vino*) in the town of Haro each year on 29 June: lariojaturismo.com

July

Celebrity Vineyards

· ·

The immaculate vineyards at Ornellaia in coastal Tuscany are the source of some aspirational Super Tuscan red wine beloved of the jet set.

In the northern hemisphere, July is a good time to make the most of the warmer weather and visit some more vineyards. Lately the concept of wine tourism has acquired a new twist, with the entry of quite a few famous names on to the winemaking scene. This transfer of talent is growing year on year as wine estates gain new owners who have no family history of farming or winegrowing.

Which comes first – the wine or the celebrity? This chicken-and-egg view of so-called 'celebrity vineyards' is intriguing, as in our era there is such an unprecedented level of interest in product endorsement. It was inevitable that wine would be caught up in the trend and, given the apparently attractive lifestyle of a *vigneron*, that a select few stars would be lured into getting involved. Whether they simply take on a vineyard that turns out good wine, or try their hands at full-scale winemaking, their famous names will do no harm to sales of bottles. Curiously enough, however, recent research by the Nielsen organization in the United States shows that wines associated with such distinguished names as Francis

Ford Coppola sell particularly well in the USA, but that in the United Kingdom there is no sign that the public is over-impressed with celebrity association, and it is just another factor when choosing one bottle over another.

One great attraction of the winemaking lifestyle is the way it is authentically linked to the earth, the seasons and the elements, far from the bright lights and artificiality of the film studio or the catwalk. One of the most famed enthusiasts for the vinous way of life (in every sense) is French actor Gérard Depardieu, who has gone on record as stating that he prefers his life as a winegrower to his other, more lucrative career in film. He unashamedly celebrates his connection to the soil: 'When we drink wine, we bond with a small piece of the earth and also with the person who made it.' His presence in the wine world is wisely 'stage-managed' by his wine mentor, Bernard Magrez (see Interview in 'September', p. 149), and his forthright personality makes a refreshing change from the mystery surrounding so many good wines. Depardieu has wine interests in

RIGHT: Fashion designer Roberto Cavalli's eye for detail extends to wine labelling.

BELOW LEFT: Cliff Richard's Vida Nova is the brilliant pink colour of hot-climate rosé.

BELOW CENTRE: New Zealand's Pinot Noir is far less well known than its Sauvignon Blanc, but actor Sam Neill is helping to change this with his Two Paddocks label.

BELOW RIGHT: Château Miraval in Provence was home to a recording studio in which Roger Waters of Pink Floyd created tracks for the album *The Wall* in 1979; this elegant rosé wine is a tribute to the band.

France; on the lovely Italian island of Pantelleria (with his former wife Carole Bouquet); and in North Africa, where he is truly at the cutting edge of winemaking, in such Muslim countries as Algeria. His film work displays a willingness to take on a challenge, and it is entertaining and impressive to watch him do the same with wine.

Other notable names from the film world who are involved in wine production include, as mentioned, the director and producer Francis Ford Coppola, famed for his series of *Godfather* films (1972–90); he is a serious winemaker, not just a dabbler, in California. Another passionate enthusiast who is supporting his local economy is the actor Sam Neill, perhaps best known for his role in *Jurassic Park* (1993). In the remote Central Otago region of New Zealand, he is producing some excellent Pinot Noir under the Two Paddocks label. Meanwhile, in the Niagara region of Canada, another actor, Dan Aykroyd (one of the original members of the Blues Brothers act), puts his name to some fine Icewine, which has won awards. In another example of someone returning to his roots, the Spanish actor Antonio Banderas has invested some of his Hollywood funds in a vineyard in Ribera del Duero. Johnny Depp likes his lifestyle in the South of France so much that he has bought a vast agricultural estate there as a gift for his partner, Vanessa Paradis.

A select band of musical performers has also taken a keen interest in wine. Olivia Newton-John, the legendary blonde star of *Grease* (1978), promotes unpretentious, enjoyable wines made in Australia under the Koala Blue label. Another famed blonde, Madonna, is happy to lend her name to wines made at her father's Ciccone Vineyards in Michigan, although she has no say in the running of the business.

RIGHT: Antonio Banderas takes a keen interest in his pristine vines in the upmarket Spanish region of Ribera del Duero.

FAR RIGHT: Unlike some more dilettante 'celebrity' vineyard owners, *Jurassic Park* star Sam Neill plays a hands-on role amid his vines.

OVERLEAF: New Zealand's Central Otago region is austerely beautiful in July – the depths of winter in the southern hemisphere. These vines are yet to be pruned.

In the scorching Algarve region of southern Portugal, the British singer Sir Cliff Richard has enlisted assistance from various expert winemakers, including Australian expatriate David Baverstock, to create some decent and drinkable wines from his estate, known as Vida Nova. Richard is known as the 'Peter Pan' of pop music for his longevity in the business; and another enduringly successful British musician is Sting, who has an enviable farm in Tuscany, where he and his wife, Trudie Styler, produce a range of honey, olive oil and red wines under the Il Palagio branding. There is a strong emphasis on an organic approach and freshness in all these products.

Tuscany is also a playground for many of Italy's famed fashion designers. Wine and fashion have a good deal in common: there are what could be described as 'haute couture' wines, handmade, rare and expensive; 'prêt-à-porter' wines, which are authentic and attractive, but more affordable; then every shade of lesser wine, right down to budget supermarket own brands, which

may look just as good as the finer versions but, like such clothes, do not last nearly as long! This analogy is not lost on such fashion leaders as Roberto Cavalli and Salvatore Ferragamo, both of whom have estates in Tuscany. Cavalli's is modestly labelled Tenuta Degli Dei ('wine of the Gods'), while Ferragamo's Il Borro is not only a wine estate but also an entire village where you can eat, sleep and live the lifestyle of a fashion baron.

Sports stars also have vinous aspirations. Ex-football player David Ginola makes award-winning Provençal pink wines, and golfers Greg Norman and Ernie Els have professional interests in wine. Perhaps the most unlikely wine endorsement comes from the American lifestyle guru Martha Stewart, who is putting her name to a selection of wines from E. & J. Gallo, the giant California wine business. Her venture is a long way from the romantic vision of a celebrity sipping wine while gazing out at misty vineyards, but does illustrate one central fact: winemaking and drinking today are no longer minority, elitist activities; they are right there on centre stage.

Wines and Big Occasions

···

Making a display of attractive glasses on a celebratory table is part of the fun, and helps to show off the particular style of each wine served.

It may be a wedding, a big family party or an annual gathering of passionate players of croquet, cricket or boules. Whatever the reason, there will be a crowd and a need to ensure that everyone has a marvellous time, no matter how old they are or what their taste in wine and food. These events are a great challenge, and many hosts abandon the struggle and simply engage the professionals – which is a safe option, but one that takes away the fun of doing things on a grand scale yourself.

Think about those foodie films you may have enjoyed: *Amarcord* (1973), *Babette's Feast* (1987), *Big Night* (1996) or even *Four Weddings and a Funeral* (1994). There is such joy in pulling together a disparate group of people and seeing them all relax over good food and wine. This pleasure principle accounts for why so many of us nurture a secret desire to own or run a restaurant,

despite all those television programmes demonstrating why this is in fact serious, hard work, day in and day out. But catering for just one day or evening – why not?

As this is July, in the northern hemisphere a marquee may be involved, and serving complex, hot dishes may be difficult. Yet there are ways to avoid the inevitable buffet. If you consider the different countries in the world of wine, there are many dishes that spring to mind that are simply ideal for quantities of people. In Italy, there is pasta of all kinds, or risotto; in Spain, paella; in Portugal, fish or meat served in a *cataplana* (a broad, domed dish); while in France there is bouillabaisse and ratatouille in Provence, or rich *daube de bœuf*. Other countries have evolved their own versions of these dishes: ostrich stew in South Africa, instead of a beef version, or, in Australia, a hot and spicy Asian-influenced variant on the theme of risotto.

Simplicity is a keynote for successful large-scale entertaining. There is something rather gloomy about elaborate buffets that feature quantities of cold food and copious mayonnaise – even the bread seems somehow frozen in time. A little interaction with the food is more fun, and family or group members can be asked (bribed?) to help serve from chafing dishes kept warm with spirit lamps. Alongside the hot food, you can offer as much salad, fruit and fancy dessert as you choose. It can all still look lovely, especially when seasonal flowers bedeck the tables.

It is perfectly possible to choose wines that are likely to keep everyone happy and not cost the earth. As with the dishes mentioned above, think of wines that are of the country, such as Vin de Pays in France, rather than top names or boutique wineries. You might want to consider a rustic touch, such as serving the wines in attractive carafes made of glass or earthenware, to create a real sense that you are sharing in a family celebration in Sicily or Seville.

Italian Style

~

To start: Fine olive oil, good bread, olives and sun-dried tomatoes on each table, and serve Prosecco freely.

Pasta: Make or buy fresh pasta. Serve with a choice of sauces for all palates, with a touch of luxury: perhaps smoked salmon with cream, mushrooms with truffle and a game *ragù*.
Wine: More Prosecco, or a white Fiano or Gavi; and red Chianti or Nero d'Avola.

Main course: Roast leg of pork (or wild boar if available) served with a selection of seasonal vegetables, the latter made into a tasty casserole (this can feed your vegetarians).
Wine: More white and red, as served with the pasta.

Dessert: Prepare fresh soft fruit by dipping them in egg white, then in sugar, and set aside to let the egg white dry. Serve a celebratory, creamy cake, alongside a beautiful display of the prepared soft fruit.
Wine: Asti Spumante or a Moscato dessert wine.

French Style

〜

To start: A selection of crudités, such as grated-carrot salad, celeriac *remoulade*, cherry tomatoes
and fresh radishes served with French bread and top-quality butter.
Wine: A good-quality sparkling wine, such as Crémant d'Alsace or Champagne.

To follow: Perfect, tiny tartlets filled with either spiced Provençal vegetables or crab with cream,
egg and a dash of hot pepper.
Wine: A fine rosé from France.

Main course: Hot casseroled *bœuf en daube*, creamy mashed potatoes and a complementary
salad that includes toasted nuts and seeds for the vegetarians, plus
a dish of mushrooms cooked with butter and garlic.
Wine: Good white and red Bordeaux, or a choice of wines from the South of France.

Dessert: A spectacular pyramid of profiteroles and glasses of Champagne or sweet Monbazillac.

Iberian Style

〜

To start: A selection of tapas, including olives, chickpeas with spinach,
home-made egg tortilla and anchovies.
Wine: Chilled Cava, dry sherry or a white wine from Galicia.

To follow: Quantities of fine Serrano ham, good bread and olive oil; white or green
fresh asparagus, with vinaigrette.
Wine: White and red Rioja.

Main course: Paella made with organic chicken, fresh mussels and other seafood of your choice,
with a good amount of saffron and a pinch of turmeric. Serve with green salad.
Wine: More Rioja.

Dessert: Orange cake, with oranges marinated in Spanish brandy and brown sugar.
Wine: Cava, or a glass of Moscatel per person.

Visiting the Vineyards of Tuscany and Umbria

The map below left shows the Italian regions of Tuscany and Umbria. Wines produced in these regions include Chianti, Vino Nobile de Montepulciano, Brunello de Montalcino, the 'Super Tuscans', Orvietto and Vin Santo dessert wine.

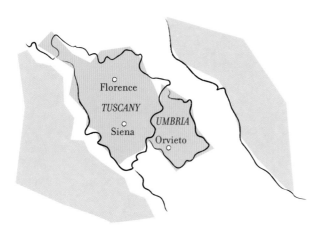

Welcome to 'Chiantishire', which for centuries has been the idyllic playground for the British upper class, and is still the place for the discriminating international visitor to rest and relax in high summer. Tuscany's hidden villas, estates, swimming pools and well-tended vineyards all seduce the eye and the mind. With such a history, the region could be said to have an unfair advantage in the world of wine tourism, yet, curiously, it retains an unspoilt charm and a feeling of timelessness. The same can be said of Umbria, where such towns as Orvieto (noted for its fine white wines) sit amid vineyards that look exactly like those painted by the Renaissance Old Masters.

Start your visit in historic Florence, a vibrant destination packed with art treasures (and a plethora of fellow tourists surging across the historic bridge over the Arno). Once your cultural hunger is sated, head out of town to Fiesole, a gloriously romantic small community with several fine hotels, including the remarkable Villa San Michele, with a façade decorated by none other than Michelangelo and views down across the Duomo of Florence. There are also some small boutique wineries to visit, where you can sample the magnificent local red wines produced from the Sangiovese grape. In summer, the weather here is glorious, yet the winters in the Tuscan hills can be harsh, and grape and olive must struggle for survival. The locals swear that this 'Tuscan Paradox' is the reason both the wine and the olive oil taste so good: they have survived tough weather and the depredations of the local wild boar and native songbirds.

For this is a place to enjoy nature in all its diversity. Your pool is likely to be invaded by giant toads in need of lifesaving; and at dusk the barn swallows swoop and dive in search of insect life.

Drive on into the Chianti countryside. For a taste of political history combined with wine tasting, stop or stay at the Villa Mangiacane, built in 1534 for the Machiavelli family; again, this small, elegant hotel has stunning views over Florence, and a landscape that is largely unspoilt, covered with vineyards and forest. Continue on to the town of Greve in Chianti, at the heart of the Classico vines. Here you could visit the Castello di Querceto, a fortified mansion with a lovely, formal garden and a range of excellent wines. You are likely to meet at least one member of the François family, which has owned the estate for more than a century; it is a curious mixture of the casual and the formal. They also offer *agriturismo*, with rooms in a farmhouse set amid the vines, affording a grandstand view of the *cinghiale* – wild boar – roaming by night.

Antinori and Ricasoli are two aristocratic names to conjure with here in Chianti; a visit to their vineyards is an experience to be savoured and relished. Apart from red wines based on Sangiovese, this region is also the source of some delicious Vin Santo, dessert wine made from grapes dried on special racks. Taste the wine made by Marco Ricasoli-Firidolfi at Rocca de Montegrossi to experience a form of time travel: his spicy, rich and haunting Vin Santo tastes just like something one might imagine being served at a medieval banquet.

If your taste runs to knightly romance, be sure to visit Siena, where medieval guilds known as *contrade* stage horse races – the famous Palio di Siena – twice a year, in July and August, in the ancient central Piazza del

FAR LEFT: The Villa Mangiacane, a luxury small hotel and winery in the heart of the Chianti region that has historic associations with the Machiavelli family, has extraordinary views from its terrace looking towards Florence.

LEFT: The wines of Ornellaia are so-called Super Tuscans, made with Merlot and Cabernet Sauvignon grapes as well as the classic local Sangiovese.

Campo. The cathedral here, striped like a marble mint
humbug, is even more impressive than the Duomo in
Florence, and has retained its character despite being
immensely popular with tourists.

If you love white wine, there is not a great deal
on offer in this part of Italy, but the Vernaccia di San
Gimignano is a delicate, fragrant example, and the town
of that name boasts some curious, ancient towers and
numerous good places to eat. Again, you will find many
fellow tourists, but the atmosphere is relaxed and
agreeable. For a town of almost absolute architectural
perfection (and some truly elegant red wines), head for
Montepulciano, perched on a hill and encircled by
ancient fortifications. The Vino Nobile of this area is not
to be confused with Montepulciano from other regions;
this is a real aristocrat, named not for the grape but for
the place. Apart from winemaking, this town is famed
for its metalwork, and a visit to a workshop here is

memorable, as are the views from any *osteria* – a simple
restaurant principally devoted to serving wine – on the
ancient ramparts as the sun sets.

Not far from here are the famed vineyards that
produce Brunello di Montalcino, one of Italy's most
expensive wines. The vineyards are on steep slopes, some
hidden among the chestnut trees, and yielding the finest
Sangiovese wines of all. To appreciate their depth and
complexity, try those made by Emilia Nardi, a passionate
enthusiast for the grape, who has her own experimental
vineyard. These are wines for long maturation in private
cellars, to be served with respect on a special occasion.

Tuscany is not all hillsides and forests; it has an
attractive sea coast, and the Maremma region is home to
several unique vineyards, such as Ornellaia, Tignanello
and Sassicaia, making top-notch wines from such classic
Bordeaux varieties as Merlot and Cabernet Sauvignon.
These Super Tuscans stormed on to the market a couple

RIGHT: The Tuscan hill town of Borgo di Pereta enjoys a lofty position near the vineyards of the Fattoria Le Pupille estate.

BELOW: The ancient Lungarotti wine cellar retains its historic press and wine storage jars similar to those used by the Romans; they topped the wine with olive oil to preserve it.

of decades ago and are now hotly sought after by collectors worldwide. There are also the delicate, suave wines of Morellino di Scansano – a name single-handedly revived by Elisabetta Geppetti (see Interview, p. 117) – which seem to have a faint fragrance of sea air about them.

To the south-east, Umbria is a huge region with much to offer in terms of art history, wine tasting and gourmet dining. Perhaps the most famous wine of the region is Orvieto, a fairly straightforward white wine. If visiting the region, seek out local red wines from such producers as Lungarotti. The Lungarotti family (see interview with Chiara Lungarotti overleaf) are dominant local wine producers who have also devoted considerable effort to attracting tourists, with a wine museum, a museum of the olive and a fine hotel and spa. In truth, you could travel to Tuscany or Umbria a dozen times and still not begin to scratch the surface of all that is to be seen, experienced and tasted in this historic area of Italy.

FAIRS AND FESTIVALS

· ·

If you choose to visit South Africa in July, try to tie in with the Stellenbosch Wine Festival in the Western Cape. Thousands of visitors gather there to taste hundreds of wines, meet winemakers and taste local cuisine: wineroute.co.za

For lovely, fairytale scenery, travel to Alsace in France and enjoy one of the seasonal wine celebrations, such as the Fête des Vins at Pfaffenheim: ronde-des-fetes.asso.fr

English wine producers are flexing their muscles these days, winning the occasional contest, even against their historic rivals the French. Sparkling wines from England are particularly impressive, but white and red table wines are also coming up fast. Travel to lovely Sussex for the annual food and wine festival at Glynde Place: glyndefoodfestival.co.uk

Chiara Lungarotti

CEO of Cantine Giorgio Lungarotti, Umbria, Italy

Cantine Giorgio Lungarotti was established by Giorgio Lungarotti in 1962.

lungarotti.it

Could you give me a little background on your family business and history?
Winegrowing was always part of the family's *azienda agraria* [agricultural estate] and after the Second World War my father, Giorgio Lungarotti, had the intuition that the future would be based on specialization and on quality. So he began to replant the old vineyards and the lands that previously grew other crops with only vines, introducing new varietals to the area. He also developed the cultural aspects by creating, along with my mother, the Wine Museum and the hospitality structures: our boutique Hotel Le Tre Vasalle and, later, the Poggio alle Vigne Agriturismo.

Were you always destined for the wine business? Where did you study?
A very important lesson I learned when I was a child is that you can do your job well only if you really like it. I am so lucky that I really like the wine business world: I grew up in the midst of it and have lived and breathed it since I was a little girl. I studied at the University of Perugia and obtained my degree there at the Agronomy Faculty with a specialization

in viticulture. Following my degree I took a number of specialization courses at the Institut d'Oenologie at the University of Bordeaux.

Give me some idea of the classic wine range you produce.
Our wines are a profound expression of the territory from which they originate, and are produced with both native and international varietals. They are subdivided according to the estates that grow the grapes with which we produce the wines. Therefore, the Tenuta di Torgiano includes wines based on the classical Umbrian varietals, for example Rubesco and Rubesco Riserva Vigna Monticchio, produced with Sangiovese and Canaiolo, and Torre di Giano with its Vigna Il Pino cru, made with Trebbiano and Grechetto. There are also wines with a more international approach, such as the Super Umbrian San Giorgio and the award-winning Chardonnay-based Aurente. The Tenuta di Montefalco, our property in Montefalco, produces the Sagrantino di Montefalco DOCG and the DOC Rosso di Montefalco; and last but not least is the Fattoria del Pomelo, which produces our young and immediately pleasant

wines, such as the varietals Pinot Grigio, Grechetto, Cabernet Sauvignon, Sangiovese and the very popular Brezza white and Castel Grifone rosé.

Do you feel the wines of Umbria are well known on world markets?
In many markets, Umbria is regarded as a new wine frontier, while in the more developed wine markets it has been recognized for many years. In the 1970s and 1980s, however, it often happened that when our wines were cited on the wine lists of the most prestigious restaurants, they would frequently be grouped under the region of Tuscany.

Can you describe some of the attractions of your estate and your region for the wine tourist?
Torgiano is definitely a wine-tourist destination, especially since it boasts such a unique attraction as the Wine Museum. This was opened to the public in 1974 and is one of the most important wine museums in the world. This private collection covers the history of wine with all its bonds to the various civilizations in which it has developed and spread. From the origins to today, it covers a time span of more than five

Elisabetta Geppetti

of Fattoria Le Pupille in the south-western Maremma district of Tuscany, Italy

. . . .

*Elisabetta Geppetti has been making wine
at her estate, Fattoria Le Pupille, since 1985.*

elisabettageppetti.com

thousand years through a wide variety of collections: archaeological finds, ethnographic items, ceramic pieces and antique publications, to name but a few.

In Torgiano there is also another museum dedicated to the other pillar of Umbrian agriculture: the Olive and Oil Museum. Both museums are managed by the Lungarotti Foundation, of which my mother, Maria Grazia, is Director.

Lastly, the latest addition to the tourist attractions we offer is represented by the vinotherapy spa called BellaUve, located within the Hotel Le Tre Vaselle, where the treatments are just … di-vine!

Finally, could you select just one of your wines and explain why it is your favourite?
Our Rubesco Riserva Vigna Monticchio is my favourite wine because of its versatility: a very full-bodied but elegant Sangiovese-based wine that is great to enjoy with braised and roasted red meats, game and aged cheeses. It's perfect even in a romantic tête-à-tête with your beloved partner. Or – why not? – just sipping it while you are seated in a comfortable armchair reading a good book in a relaxing moment.

Tell me a bit about how you restored the Morellino di Scansano Denominazione di Origine Controllata (DOC).
Twenty-five years ago, the DOC of Morellino di Scansano was just beginning when my story began with it. I like to consider the few of us who brought Morellino di Scansano to the level that it is now as pioneers. Even when I was young, I always believed in this land, in its value and potential. In the 1980s I was among the founders of the Consorzio Tutela Morellino di Scansano [a group that revived this ancient wine region], and in the first years I was the president. It was the beginning of a new birth for these Italian wines.

You have a busy family life and you also travel frequently. How do you achieve a work/life balance?
At the beginning it was very difficult indeed. I was in conflict with myself: I felt divided between my children and my passion. I let my children participate in the life of Le Pupille and I can proudly say that they are enthusiastic. They participate in every way, not only in the official occasions, but also in the

everyday work: Clara helped me in choosing a label, Ettore joined in the harvest, and so on.

Tell me why a visitor would enjoy a holiday in your part of the Maremma.
Maremma is a fascinating land, rich in traditions. From the sea to the mountains, it has been the cradle of the Etruscans. You can find art and history, uncontaminated beaches and seas, natural parks and oases. In the last fifteen years it has also been possible to find restaurants where you can drink some of the best wines of Italy and eat very good and healthy food.

What plans do you have for Le Pupille in the future?
Work, work, work! Work with the same passion and dedication, trying as ever to achieve the best quality, hoping for good luck with the weather.

Finally, could you nominate just one of your wines as a favourite, and say why?
Saffredi. It represents the wine of my youth and it is dedicated to the person, Fredi Gentili, who helped me to become the person that I am today.

August

Unusual or Forgotten Vineyard Regions

Opposite: Gentle hills roll off into the distance above the Sokol Blosser winery in Oregon's Willamette Valley.

Below: Visitors enjoy the views at the Camel Valley vineyard, Cornwall, famed for its wine made with the Bacchus grape.

Now that the world seems a smaller place, and it can be just a matter of a few clicks on the computer mouse to book an air ticket to Timbuktu or Mandalay, it is time to rethink the wine map. As we have seen in previous chapters, wine tourism is gaining in popularity each year, and August is the month when many of us escape working life. Beyond the classic regions of Europe, such as Bordeaux or Tuscany, and the notable New World zones, such as the US Napa Valley or the Australian Barossa, there is much more happening out there, and it is startling to assess how widespread vine cultivation has become. Naturally, some of these planting experiments are not going to rival a Grand Cru any day soon, but it is exciting to contemplate how the production of wine is growing, and how much concomitant interest there is throughout the world in the pairing of food with wine.

Europe

First of all, do not forget the countries that are significant and historic producers but are now so often neglected by consumers. These include Portugal, home of some fantastic, modern rosé, some excellent, light Vinho Verde and deep, rich reds, from the hot Douro Valley and farther south in the Ribatejo and Alentejo. Local, traditional grape varieties, such as the chocolate-scented Trincadeira, add unique characteristics to the wines. This country is also home to magnificent cork-oak forests, where wild flowers bloom freely in the spring and eagles soar above.

Austria and Switzerland are also frequently overlooked by those who enjoy light, refreshing white wines and crisp, tart reds. Only the fortunate few who ski, hike or take a spa cure chance across these delicious, low-alcohol wines. Look out especially for the versatile white Grüner Veltliner in Austria, and some remarkably good Swiss Merlot.

Luxembourg has some serious wine drinkers, among the thirstiest in Europe, which might account for rare sightings of these wines on the market. Along with some agreeable sparkling wines from such producers as Caves St Martin and Domaine Alice Hartmann, the country also produces refreshing Riesling, Pinot Gris and Pinot Noir.

Bulgaria and Romania had a moment in the sun in the 1970s, when their wines became popular, but since then they have struggled for a share of the wine market. Romania has a very long history of grape growing; Bulgaria had Roman vineyards but returned to production only in the Soviet era. Overall, in both countries, the red wines tend to be superior to the whites. Look out for Bessa Valley Bulgarian wines, or Cabernet Sauvignon from the Damianitza winery. In Romania, Prahova Valley is a reliable label to look for; the winery is within a day's drive of Bucharest. Georgia and Moldova have chequered histories, coloured by the fact that they were Soviet territories for many years, and corruption has since been rife. Now, at last, they seem to be emerging from this cloud, and there are some good wines out there, notably the tongue-twisting Rkatsateli, a creamy, light white, which is also grown with great success by Dr Konstantin Frank in New York State (see 'October', pp. 160–61).

Hungary produces one of the world's great dessert wines – Tokay, as beloved by connoisseurs today as it was by the Russian tsars of old. Apart from the well-known brand Bull's Blood, there are also some excellent red wines here. Look out for the Szeremley label, owned by a dynamic entrepreneur; the Figula winery near lake Balaton is also a source of good Cabernet.

Greece is now becoming quite a connoisseur choice, as modern, young winemakers experiment with ancient grape varieties. Even the pungent, pine-scented Retsina is now being made with a really fresh twist. Try a red from Tsantali, produced on the slopes of Mount Athos, or a Greek Chardonnay for a really new take on this familiar grape.

Croatia, Slovenia and Macedonia were all part of former Yugoslavia and at that stage of history were not really renowned for their fine wines (on the contrary, Laski Riesling was a byword for dull white wine). Yet, between the steep, northerly slopes of Slovenia, with its Austrian-style whites, and Croatia, with such deep, dark reds as Dingač, there is huge potential for excellence here. Now that these countries are at peace, and investment is flowing in, it is only a matter of time before some truly fine wines emerge on to the market.

LEFT: British seafood chef Rick Stein (centre) poses with his Cornish winemaker friends at the Camel Valley vineyard.

LEFT, CENTRE: The wines of Croatia are attracting increasing critical acclaim and foreign investment.

LEFT, BOTTOM: This bucolic scene in the Golan Heights demonstrates that wine production is now a serious part of the agricultural scene in the Middle East.

BELOW, CLOCKWISE FROM TOP LEFT: Gobelsburger Grüner Veltliner from Austria, a delicate dry white wine; fine red wine from top Lebanese producer Château Musar; that perennial Greek classic, Retsina; quality red table wine from the port vineyards of the Douro, Portugal.

Meanwhile, enjoy the tourist potential, which is also still relatively untapped, and sample some intriguing wines on the spot. One winery attracting international attention is Grgic Vina in Croatia. In Macedonia, the new Château Kamnik winery, close to Skopje, has ambitious plans to make high-quality wine. By contrast, Vino Kupljen in Slovenia has a family history dating back to 1836 and runs its own wine school.

The United Kingdom is fast gaining a reputation for its excellent sparkling wines, to the extent that some Champagne houses are looking to establish vineyards on the green hills of Sussex and Hampshire. Vineyards thrive in Wales as well as in England, but the uncertain climate is still the main reason why wines from these regions are unpredictable; at their best, in a good year, even the reds (notably some good Pinot Noirs) can win awards. Top wineries incude Camel Valley, a Cornish venture that produces superb white wine from the Bacchus grape; Denbies, an entertaining day out near London; and Nyetimber, which produces fine sparkling wine in West Sussex.

Twenty years ago, Cyprus had a reputation as a source of sweet, sticky 'sherry' and a rich wine called Commanderia, made with dried grapes. Now this country is fast modernizing its vineyards, and producers are improving the quality of the wines they are offering to the world. The very best are from the KEO winery; other good labels include Sodap and K&K Vasilikon, a lively family winery not far from the resort of Paphos.

LEFT AND BELOW: Vines have been cultivated in the twenty-eight villages of the Tokay district of Hungary for centuries. The distinctive dessert wine is produced from Furmint, Hárslevelü and Muscat Ottonel grapes.

The Middle East and Africa

....

Israel was long associated with sweet wines, many of them kosher wines for a specialist market. However, like Cyprus, the country is making great strides to try to overcome the disadvantages of its hot climate to make modern wines, some (controversially) from the Golan Heights, where conditions are very favourable for wine-grape cultivation.

Lebanon has been winning awards for its wine for many decades, and the French influence here has survived conflict and economic hardship to persist in the realization of some superb Cabernet Sauvignon and other classic varietal red wines. Châteaux Kefraya and Château Musar, both based in the Bekaa Valley, are influential names to watch, and Massaya, also in the valley, is a modern producer with organic principles.

Algeria, Morocco and Tunisia were all producers of wine during the time of the Phoenicians, some 6000 years ago (as was Lebanon); later, the Romans planted vines in these countries. During the nineteenth and early twentieth centuries, the French colonists used the deep-coloured wines from these regions to enrich their table wines back home. Today Algeria is a difficult culture for winemaking, but Gérard Depardieu (see 'July', pp. 103–104) and Bernard Magrez (see Interview in 'September', p. 149) are attempting the task. They are also busy in Morocco, where the wine-exporting culture had hitherto virtually vanished. There are now wineries in the foothills of the Atlas mountains, including Les Celliers de Meknès, a reliable brand. The Magrez/ Depardieu benchmark red, made with Syrah and Grenache, is called Lumière de l'Atlas.

Asia

....

India is a vast country that consumes only a tiny amount of wine at present, yet it has some excellent producers and there is scope for substantial, significant improvement. French grape varieties are now being grown, where once all wines were made with table grapes; Grover Vineyards in Bangalore is leading the way. Working with French wine guru Michel Rolland, this winery has already won acclaim from the critics for its Cabernet Sauvignon-Shiraz blend.

Thailand really does not have the ideal climate for *Vitis vinifera*. Grapes are averse to tropical humidity: it encourages leaf growth over fruit and can mean a high risk of fungal disease. Nothing daunted, such wineries as Monsoon Valley (a division of Siam winery), south-west of Bangkok, are making wines that are intended to complement the complex flavours of lemongrass, coriander and curry, which dominate the local cuisine. This involves a whole new approach to winemaking, which is being closely observed by the rest of Asia.

China is already in sixth place for overall world wine production and, where once rice wine ruled, the Chinese are now developing a taste for the grape. In ancient times, there seems little doubt that vineyards thrived here, but there was then a gap of thousands of years before wineries reappeared recently, led by China Great Wall Wine Co., Ltd in 1983. The harsh winters across much of China mean tough growing conditions for classic *Vitis vinifera* grapes, but modern techniques are helping to make vineyards more viable. The Chinese generally like a sweetness in their wine and, as in Thailand, this style complements their spicy and often oily cuisine. Look out for quality wines under the Château Changyu-Castel label.

Wine with Barbecues

Even the most stalwart vegetarian tends to admit to a fondness for barbecues; there is something about the snap and sizzle, and the scent of woodsmoke, that fires up the taste buds. Many wine-producing areas are devotees of the grill too, often using vine cuttings as fuel. These delicate twigs have the most delectable aroma when set alight. In Bordeaux, sample *pré salé* ('salty fields') locally reared lamb cooked in this way; or in Argentina relish the quality of their barbecued beef.

From the *bistecca alla fiorentina* of Tuscany to the whole roast suckling pig of Rioja, there is a lengthy tradition of pairing the very finest of wines with meat or fish cooked over an open fire – although the very act of cooking in this way is quite a challenge to wine's aromas. If your main dish has a crispy carbonized crust, this could easily annihilate a light, fruity wine. So you need to seek out wines with real muscle and tannins, which will withstand the onslaught and create an appealing balance on the palate.

Certain grape types are natural partners to the barbecue: spicy Syrah/Shiraz; deep, dark Nebbiolo and Barbera; Grenache/Garnacha, Carignan (or Cariñena) and Mourvèdre; and a raft of other Mediterranean red varieties, including Primitivo, Nero d'Avola and Aglianico. In the New World there is pungent Pinotage from South Africa; chunky Zinfandel from California; Malbec from Argentina; Petit Verdot from Chile; or Tannat from Uruguay. Many of these wines have fairly high alcohol levels, and the sweetness derived from the alcoholic content is masked by the earthiness of the grilled flavours, so beware drinking in the full sun.

Many of these wines benefit a great deal from being opened ahead of time, to aerate and soften the tannins. So, to release their full richness and depth, consider opening a few bottles for your guests in advance.

Can a white wine or a rosé ever be appropriate for a barbecue? Absolutely – particularly if you are serving rich, oily tuna, salmon or swordfish, all of which grill superbly and form a very uneasy alliance with red wine. Pink wines, even those with a hint of sweetness, make an intriguing contrast to lighter grilled meats and fish, while some red grapes, such as Gamay, with their joyous jammy fruit, sing with seafood.

For your next barbecue, try some of the wine-pairing ideas on the facing page.

Food	Wine
Beef	Hamburgers seem natural allies with Syrah/Shiraz; steak is great with Pinotage, and Brunello di Montalcino (a substantial, rich red made with Sangiovese in Tuscany); lively, salty Tannat from Uruguay.
Chicken	White ('blush') Zinfandel is surprisingly delicious, as the sweetness complements the chicken well; red wine from Catalonia, such as Sangre de Toro from Torres.
Grilled vegetable kebabs	Zinfandel from California, or Primitivo from Italy (these grapes are said to be cousins); Côtes du Rhône; Argentine Malbec.
Lamb	Classically paired with Cabernet Sauvignon, but choose one that is a full, rich version from California, Australia or Chile rather than a light, tart Bordeaux claret; Priorat or Ribera del Duero red from Spain; Greek red wine.
Mushrooms	Any wine made with the Barbera grape; or bring out the mushrooms' sweet nuttiness with a white wine, such as Jurançon from the South of France, or mature New Zealand Sauvignon Blanc.
Pork, sausages	Barolo or Barbaresco from Piedmont are great natural partners for these foods; red wines from the Douro Valley in Portugal; Australian Shiraz.
Salmon, swordfish, tuna	California Chardonnay; Chilean Sauvignon Blanc; Grillo from Sicily; Vinho Verde from Portugal; dry pink wines.
Scallops wrapped in bacon	Australian sparkling Shiraz or dry Semillon; California Fumé Blanc; white Rioja.
Shrimp, prawns, crayfish	Deep-pink wines made with Syrah/Shiraz or Grenache from Spain, Chile or Provence; Chenin Blanc from South Africa; aged Beaujolais.

Visiting the Vineyards of California and the Pacific Northwest

The map below left shows the American states of California, Oregon and Washington. The highlighting shows the area's main wine-production regions.

In the decades since the famed 'Judgement of Paris' of 1976 (a wine-tasting competition), when California wines comprehensively outclassed some of France's finest bottles, confidence has steadily grown and the wine industry of the American West has become a class act. The vast majority (some 95 per cent) of the wine produced in the United States comes from California, with some very attractive offerings from nearby Oregon and Washington State. All of this wonderful region is worth exploring; from the smart boutique wineries of Napa to the folksy charm of Oregon's Pinot Noir heartland, there is so much to enjoy – and to taste.

Start your journey in Santa Barbara County, in the warm southern vineyards, which were the setting for the wine-tasting road-trip film *Sideways* (2004). This is a lively, experimental region with some excellent red wines made from Bordeaux grape varieties, including Merlot

RIGHT: The steep-sided wine valleys of California are baked dry by the sun in summer, but winter rains ensure that the vine can thrive here and produce sensational wines.

RIGHT, CENTRE: Walla Walla County in chilly Washington State is now home to some excellent wines, and new vineyards are being created all the time.

RIGHT, BOTTOM: In many corners of California and the Pacific Northwest, such traditional fruit crops as pears, apricots and plums are still grown alongside grapes.

and Cabernet Sauvignon. A visit to the Blackjack Ranch Winery in Solvang, originally financed by the owner's invention of the California system of blackjack for casinos, can be amusing. Sip fine red wines at the bar, which was once a lane in a bowling alley. There is a frontier feeling everywhere here; the industry began only in the 1990s. Since then wine producers have battled insect infestation and other problems, but there is a tangible air of optimism. Some critics compare the climate of the region and the style of these red wines to those of Tuscany; the area certainly has both scenic beauty and great potential for creating fine wine.

Farther north, amid the redwoods above the vibrant university town of Santa Cruz, is the legendary Bonny Doon winery run by Randall Grahm, a keen enthusiast for Italian wines and those from the French Rhône Valley. His labels are unconventional, his wines

BELOW AND BELOW RIGHT: California wine labels vary from the 'wacky' to the classic, as these two examples demonstrate.

BOTTOM: Machine harvesting was pioneered in California for the vast acreages of vines there.

OPPOSITE, CLOCKWISE FROM TOP: Morning fog in Mendocino County, California; a modern buggy in California replaces the traditional tractor; the landscape of the Pacific Northwest; Château Ste Michelle, a pioneer of wine excellence in Washington State.

are intriguing, and it is worth the wander in the forest to find him. Drive onwards towards the city of San Francisco and you will encounter more wine pioneers. The founders of Ridge Vineyards were Stanford Research Institute engineers, and both the Cabernet Sauvignon and Zinfandel produced here have a cult following. Winemaker Paul Draper has had a fine reputation here since 1969. Again, it is a challenging drive but a worthwhile effort to get up the mountain to the winery, to taste and discuss these outstanding bottles. Another fascinating pioneer, this time working with Pinot Noir, is Josh Jensen at Calera Wine Company, who has succeeded in his dream of making wines to rival those of Burgundy.

From San Francisco, cross the Golden Gate Bridge and head north through the county of Marin, with its quaint seashore communities, into the heartland of the California vine – Napa and Sonoma counties. Wine grapes followed this route too, first planted from south to north by Spanish missionaries as they travelled. None of the wines was of notable quality until the eccentric adventurer 'Count' Agosthon Haraszthy brought cuttings of *vinifera* vines from Europe in the nineteenth century. Early vineyards, such as Buena Vista and Beaulieu, thrived, and were winning medals in France during the Belle Epoque, the twenty-five years before the First World War. But the onset of Prohibition ended this trajectory, and it took another forty years for the industry to recover, moving from making dessert wines and growing table grapes back to the sophistication of fine-wine production. It is intriguing to visit such a winery as Beringer, in St Helena, Napa, which has a lengthy history to recount; contrast this with somewhere entirely of the moment, such as the Hess Collection Winery, which boasts some very fine wines and a spectacular art display; or sip a fine glass of classic fizz at Mumm Napa (owned by the French Champagne house) or historic Schramsberg, an early arrival on the sparkling-wine scene.

The steep-sided Napa Valley is spectacular, but nearby Sonoma is also charming, with its feel of old-world Americana, little farmsteads and open fields. There are many welcoming wineries here, including Sebastiani and also Davis Bynum, passionate advocates for the organic way of life. Farther north again, the Russian River Valley is a great place to taste wine and enjoy lovely scenery, including redwood groves. Then there is Mendocino, with such communities as Bolinas, which is renowned for eccentricity. Appropriately enough, the top wineries here are Fetzer and Bonterra, both examples of how organic wines can also be commercial (see interview with Bob Blue overleaf).

Despite its proximity to California, the Pacific Northwest likes to stand apart from its southern neighbour; Oregon and Washington State like to tell their own wine stories. As with Napa, the prestige of Oregon's Willamette Valley can be dated from its success in competition with the French in the 1970s. Today, such classic wineries as Adelsheim produce excellent Riesling and Pinot Noir, and they have been joined by a French contingent including Domaine Drouhin, the staff of which know a thing or two about Pinot Noir. There is a degree of passion here, and an intimate feel to the wineries, which echo the experience of visiting the Côte d'Or in France (see 'November', pp. 178–79).

Farther north, there are some four hundred wineries in Washington State, apparently undeterred by the fact that this is a region where the climate makes grape growing a real challenge. This is a glorious area to visit during the harvest, bringing to mind the chilly charms of Germany or northern Italy. The resemblance

makes some sense when one knows that the huge estate Château Ste Michelle has in the past collaborated with Antinori of Italy and Ernst Loosen of Germany. Smaller producers, such as Andrew Will, make interesting wines, again with a nod to classic European regions, in Will's case Bordeaux. Throughout the region the quality standards are improving every year.

Bob Blue

Winemaker at organic winery Bonterra in Mendocino County, California, USA

Bonterra has been growing grapes organically since 1987.

bonterra.com

Could you give me a little personal history, please?

I grew up in rural Lake County in northern California – back then a community known more for growing pears and walnuts than wine grapes. I developed an appreciation of and interest in wine when I lived in Germany for two years as a member of the US Army stationed in Nuremburg. Wine was never in my family experience – in fact, we were teetotallers! Of course, the first wines I learned to appreciate were the German Rieslings, which was a great place to start. When I returned to California, I happened on a television programme about the UC Davis enology programme.

How did you come to work with Bonterra?

When I graduated from college in 1983, I started at a winery called San Martin as a lab technician. I then had the opportunity to join Fetzer. In 1987 the Fetzer family made the decision to start farming organically, and in 1990 the first official organic grapes were crushed and made into wine there. Fetzer released its first organically grown wine from that vintage in 1992. The next

vintage was called Fetzer V. Bonterra organic, and then in 1994 Bonterra was launched as its own brand. I made the very first organic red wine, a red table wine and then later the Cabernet Sauvignon, Sangiovese and Syrah. And by 1996 I was named the winemaker, making all the wines.

Could you summarize the relationship of Bonterra wines to food – I understand wine pairing is a keynote of your philosophy?

Our wines and our initiative to grow grapes organically were inspired by an organic garden back in the 1980s and 1990s. I had the opportunity to experience the work of many chefs, who were harvesting from the garden and creating an array of dishes inspired by many cultures, wrapped up in the California cuisine style. The whole notion of organic, to me, is freshness and bright flavours, and I think the Bonterra white wines reflect those qualities. They emphasize the best qualities of the grapes – whether Sauvignon Blanc, Viognier or Chardonnay – and are balanced nicely with acid, so that they are not heavily alcoholic, sweet or over-oaked. My goal

with the red wines is to capture a balance between the Old World and the New World wines. They show some restraint; a harder thing to do in making wines, it seems, these days. In the end I believe there is an elegance that is expressed in the Bonterra red wines.

How difficult is it to explain organic wines to the consumer?

Well, it is 200 per cent easier than it was fifteen years ago. I think consumers in general understand that organic is an alternative farming system; one that is easier on the environment, that prohibits genetically modified organisms (GMOs), and that reduces the use of synthetic chemicals. I think today people recognize that wines made from organic grapes are an added benefit. But first (and this has always been the case with us), the wine must be good.

Could you tell me in a few words why Mendocino and your winery are an essential stop on any wine route?

Of the northern California wine regions, Mendocino County has one of the most dramatic landscapes: located along the Pacific Ocean, it is positioned between the great north-west of redwoods and a wet, cooler climate, and, in the south, the Mediterranean-style climate of drier, warm weather. This gives it elements of both types of climate. Mendocino is mountainous and wild, with deep valleys cutting through the landscape. The vineyards are situated both in the valleys and on the slopes. Mendocino has a great diversity in wines, from Pinot Noirs and Alsatian varieties grown in the Anderson Valley to rich Zinfandels and Rhône varieties from the Redwood, Ukiah and Sanel valleys. I wish I could host the public at our winery, but currently it is open only to the wine trade. There is a great wine trail if one travels north from San Francisco to the town of Mendocino on the Pacific coast.

How about future plans? Where does Bonterra go from here?
We have several new ventures. This vintage, we are working on two new wines – a Pinot Noir and a white Riesling. The Pinot is very exciting for me; I haven't worked with the grape since 1992, and we have some nice wines just finishing fermentation and going into barrel for malolactic fermentation. The Riesling is also a lot of fun because

it goes back to where I first discovered wine. Our other venture is biodynamic wine. In 2002 we introduced 'The McNab', a Bordeaux-inspired blend holistically fashioned [according to the biodynamic principles of Rudolf Steiner] at the McNab Ranch, which is planted mostly to Merlot, Cab and some old Petite Sirah (the California twist) [Petite Sirah is cultivated only in California]. We are soon going to add another wine, called The Butler, from our really spectacular mountain Butler Ranch. It is inspired by the Rhône and is made from Syrah, Grenache, Mourvèdre, Petite Sirah and Zinfandel. After that we plan to introduce a white blend from a new ranch along the Russian river.

Finally, would you nominate a favourite wine among your selection?
I'm really drinking a lot of Bonterra Rosé right now. It's the second vintage, it's made primarily from Sangiovese, Grenache, Zinfandel and a little Syrah and Carignana, it's loaded with berry fruit and is dry. The other wine that is very special is 'The McNab' blend, which has great mouth feel and flavour.

FAIRS AND FESTIVALS
....

In August, enjoy some music while you explore California's Napa Valley wineries, during the three-week Music in the Vineyards chamber music festival, held in estates throughout the region: napavalleymusic.com

The city of Worms in south-west Germany, famous for Liebfraumilch and sparkling Sekt, has a big wine festival in late August and early September: worms.de

In France, you could celebrate the fine wines of Touraine in the Loire Valley at the Vouvray Wine Festival, featuring white wines that reach a great age in limestone cellars: vins-vouvray.com

Or, still in France, go south-west to Madiran and sample the excellent red wines of this type at the Fête du Madiran: commune-madiran.com

A very smart wine-tasting event is held at the Italian ski resort of Cortina d'Ampezzo, where top producers showcase their fine wines at Vino VIP Cortina: vinovipcortina.it

September

How Wine is Made

This is a momentous month in the vineyard. In the southern hemisphere, the important work of ploughing, essential for keeping weeds in check, gets under way. Throughout the northern hemisphere, meanwhile, the harvest is in full swing, and everywhere there is an emphasis on protecting the ripening grapes from the ravages of weather and pests. Both excess sun and sudden rain can bode ill for the winegrower: scorching heat literally burns the grapes on the vine, turning grapes to raisins in short order; while at harvest, rain is an enemy, as it dilutes the flavour and quality of slowly ripened wine grapes. It may also trigger mould growth, which can be disastrous. Weather forecasts and constant testing of sample bunches in the vineyard are vital to a successful vintage.

In many classic wine regions of Europe, there were once formal dates when the harvest was fixed to commence. In France, this was known as the Ban de Vendange. The tradition began in medieval times in an attempt to ensure decent wine quality across a region. Nowadays

OPPOSITE: The vine leaves in this Bordeaux vineyard show a hint of autumnal colour as the time for harvest approaches. After the harvest is a lovely time to see the region, as winemakers have more time and the vines turn shades of orange, purple and brown.

RIGHT: In most parts of the Bordeaux region the pickers still live in villages in the area, unlike the migrant workers in many other wine regions of the world.

FAR LEFT: Great skill is needed in cutting each ripe bunch of grapes to avoid skin damage.

LEFT, TOP: In South Africa, the coming of the 'Rainbow Nation' means that many vineyards are now under black ownership.

LEFT, BOTTOM: As every gardener knows, the close attention of wasps to fruit means just one thing: these grapes are fully ripe.

wine laws, such as the French Appellation Contrôlée (AC) or the Italian Denominazione di Origine Controllata (DOC), are used to influence each producer's decision about when to pick. Yet there is such a mass of microclimates within every wine region, that it is really a question of logical agriculture based on local expertise. Each grower who is in daily contact with his or her vines will know when the perfect moment for picking has come.

Once sugar and acidity within the grapes are in perfect balance, it is time to call in the teams of pickers – local or itinerant workers waiting in the wings to start the harvest. Alternatively, this may be the moment to ensure access to a precious machine harvester, if the vineyard is designed to permit this style of picking. There is a lot of snootiness about picking by machine. Certainly, early prototypes did leave far too many leaves in the hopper with the grapes, and could damage some bunches. But today's huge harvesters are treasured, especially in extreme climates, as they get the work of picking over and done with so rapidly, and the grapes are transferred at top speed and in peak condition to the winery. A mechanical harvester is often shared between many small-scale winegrowers.

Grapes are often transported under a layer of inert gas (such as carbon dioxide) to protect them from harmful oxidation. The gas forms a shield over newly harvested grapes within the 'gondolas' used for transportation. Otherwise, since wine yeasts live on the skins, nature has a way of starting fermentation before the winemaker has a chance to take charge. On arrival at the winery, the bunches are frequently sorted to remove any damaged goods, then weighed, and a few are tested to ensure that they contain sufficient sugar to ensure a decent level of alcohol after fermentation (between 10% and 14% by volume, depending on the local style and wine-quality level). If the grapes come from individual producers to a co-operative or a giant wine 'factory', then each grower is compensated according to the quality, quantity and ripeness of his or her grapes on arrival at the winery.

Making White Wines

A white wine is generally produced from the juice of the grape only. The grapes themselves need not be white: fine Champagne, for instance, is made from the juice of (white) Chardonnay and (red) Pinot Noir and Pinot Meunier.

The juice is separated from the skins by using a grape press. At one time, these were rustic wooden contraptions with heavy weights to crush out the juice, but today there are many and various mechanical aids, with pneumatic presses gently squeezing out the juice without adding any undesirable bitterness from squashed grape pips or skins. Temperature is absolutely key to keeping that elusive 'grapey' charm needed for most white wine. Most is sold just months after production, and it needs to taste as fresh as possible. Fermentation in giant, temperature-controlled stainless-steel tanks may not be very tourist friendly, but

it does achieve the desired flavour and allows the process to be controlled. An ideal temperature at this stage would be 10°C–18°C (50°F–64°F). The process takes two to three weeks, depending on the conditions in the winery.

Traditionally, wine fermentation took place in warm conditions, and raced away until suitable alcohol levels were reached. Then the wine would be 'racked' – transferred to a new barrel or tank – and this would halt fermentation. Over the winter months, the cellar would cool naturally, and the wine would rest and gently precipitate any residue to the bottom of the barrel. This quaint, rustic way of allowing wine to make itself is very rarely followed now, even by the most devoted of organic or biodynamic producers. The margin for error and sudden disaster is too great.

Some winemakers do use the naturally occurring wild yeasts from the grape skins to start fermentation, but most add yeasts made to laboratory standards, to ensure consistent quality in the wines.

RIGHT: In every wine region, there is huge manual labour involved in getting grapes to the press; the coming of machines to the vineyard does not eliminate human effort.

FAR RIGHT: Stainless-steel fermentation tanks are now an industry standard for controlling temperature and maintaining hygiene. This is particularly the case in the making of white wine, as the delicate fruit aromas of these grapes may be lost if the fermentation process takes place too rapidly or at very high temperatures.

Making Red Wines

For most red wines, the stems of the grape bunches are removed, as they tend to add a note of harshness in the finished wine. Then the grapes are gently crushed, not pressed as for white wines – that happens at a later stage for reds. The resulting grape mush is transferred to fermenting vats and, again, temperature is important. The vats are frequently open-topped to allow 'punching down' of the cap of skins that forms as the fermentation bubbles away. This residue of skins lends both depth of flavour and colour to the finished wine. Large-scale producers have handy machines and lots of computers to do all this work for them, but human observation is still vital.

As with white wines, the grape sugars turn to alcohol via the action of yeasts. Close monitoring ensures the alcohol level does not rocket too high; in general, a table wine should not exceed 15% alcohol, but there are now plenty of wines that are very close to this level. One way of adding more colour to a light-red wine is to press the grapes after fermentation finishes, and then add in the 'press wine'. For fine wines, this process is carefully regulated by law and local custom.

Once the red wine has been racked off from its vat, it may be selected to be aged in wood – usually oak, and frequently French (see 'October', pp. 151–53, for more about wine ageing). There is almost no wastage in a winery. Press wine is used to make a simple table wine, or distilled down for industrial alcohol in a nearby refinery. The mass of wrung-out grape skins, called pomace, makes a fine fertilizer.

OPPOSITE, TOP: Cement tanks might be unromantic, but, glass-lined, they last for decades and add no flavours to the resulting wine.

OPPOSITE, BOTTOM: New oak such as this would be used very rarely for fermentation, as it can overpower the fresh aromas of a new wine; winemakers choose it to age finer whites and reds.

RIGHT AND BELOW: Workers 'punch down' the cap of skins on red-wine fermentation tanks to keep it in contact with the fermenting wine. Oak fermentation tanks add just a hint of flavour and wood tannin to the ensuing wine.

Matching Wine with Vegetarian Food

Mushrooms in all their varied shapes and sizes are an invaluable standby for the vegetarian cook; if they are out of season, try the dried version.

There are many good reasons to eat a vegetarian diet, from the fact that grazing animals contribute to climate change, to simply appreciating the traditional cuisine of certain cultures that feature vegetarian dishes. Cutting down on red meat may well be beneficial to your health, and there is a trend towards 'meatless Mondays' as a way of incorporating a vegetarian diet into your life – the concept is intended to reduce consumption of animal fats, improve health and help the planet. It is not difficult to pair wine with vegetarian menus.

The month of September has to be the perfect moment, at least in the northern hemisphere, to sample what nature's bounty has to offer. From squash to chanterelle mushrooms, russet apples to radicchio (red chicory), this is a season for relishing fresh vegetables and fruit. First among ingredients comes an astonishing array of mushrooms, including the versatile, juicy ceps (*porcini* in Italian) and simple field mushrooms picked

and eaten the same day. Even the giant puffball can be tasty if cooked with plenty of garlic and butter. Add freshly chopped parsley, and serve on lightly toasted bread for the perfect lunch or light supper. This is really the very best way to showcase any fresh fungi, but a dash of white or red wine mixed with the mushrooms in the pan is another idea that makes it a little more special. Again, serve just with bread and a matching wine.

The earthiness of fungi and root vegetables is a natural partner for rustic red wines, such as California Zinfandel, Italian Primitivo and Barbaresco, or Spanish Priorat. When a delicate dish of chanterelles or oyster mushrooms is on offer, however, think of a ripe, rich white wine made with the Chardonnay, Viognier or Pinot Gris grape. Alsace is a great region for finding good, full white wines that have enough structure to balance the rich flavours of fresh fungi; look for Pinot Gris or dry Muscat.

Californian Zinfandel is a great partner to vegetarian cuisine, such as stuffed courgette flowers and the aubergine dishes that are a staple of Middle Eastern meze.

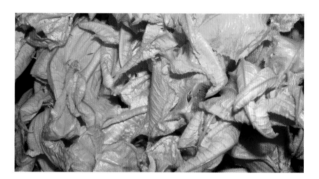

Some vegetarian dishes can be markedly sweet, incorporating carrots, onions, squash or pumpkin. These need a counterbalance, and it is an excellent idea to choose oak-aged red or white wines. The slightly harsh, tannic bite of such a wine helps to make the tinge of sweetness in, say, pumpkin ravioli, taste even more toothsome. Wines to consider include full-bodied Australian Shiraz with vegetable casseroles; young red Burgundy with onion-based dishes; Cabernet Sauvignon from the Languedoc region of France with garlicky ratatouille; and oaky California Chardonnay with vegetarian tarts or pizza.

Vegetarians value nuts and such legumes as beans and peas to add protein to their diet. September is the start of the season for wonderful fresh cobnuts, hazelnuts, walnuts and almonds. There are complex flavours at work here, particularly when the nuts are lightly toasted before serving. Once again, consider oak-aged wines to create a rich, nuanced flavour combination. A walnut sauce on pasta is a marvellous match for barrel-aged Chianti; gourmet salad with radicchio and toasted nuts would be excellent with California Cabernet Sauvignon; and a Mexican-inspired creamy mole sauce with nuts and chocolate would be well complemented by a young, full-flavoured white or red Rioja.

Mixing beans, peas or dried legumes with herbs makes a dish that is very tasty when accompanied by a rich white wine. A vegetarian version of classic French cassoulet matches very well with Viognier, Semillon or South African Chenin Blanc. Butter beans and cannellini are great for cold antipasti dishes and, again, go well with full white wines, such as New World Chardonnay and Semillon.

The olive harvest comes later in the season, and these delicious, healthy additions to the diet can dominate dishes with their heady, tangy aromatic flavour.

Beans in all their colourful varieties (these are *fagioli* in a Calabrian market) are an essential of the vegetarian diet and a great source of protein.

They are often combined with tomatoes, which are also awkward – high in acidity, which may clash with wine – and sometimes with artichokes, which, as we have seen (p. 65), are almost anathema to wine. But it is entirely possible to find a suitable wine to match these Mediterranean vegetables. First of all, consider cutting the sharpness of tomato dishes by adding just a touch of sugar when cooking. Combine artichokes with mushrooms and onions to reduce their oily richness. Chilli pepper, used judiciously, can also balance these assertive flavours. Try adding a dash of Tabasco or a sprinkle of pepper to your next vegetarian pizza or *pissaladière* (the French version, made with shortcrust pastry). Then serve with a Syrah from the South of France or from South America.

Green olives and almonds have a natural affinity with white wine, however, and are delicious with Verdicchio or Fiano from Italy, or New Zealand Sauvignon Blanc. These wines all have a slightly herbal quality on the nose, which complements the 'grassy' flavour of the olives. Green peppers are excellent with Sauvignon Blanc as well. Try lightly frying them in olive oil, adding chopped green olives and then scrambled egg. This brunch would be superb with a California Fumé Blanc or perhaps a glass of sparkling wine from the Loire Valley.

There are no strict rules for pairing wines with vegetarian cuisine: it is a great arena for experimentation. When in doubt, consider choosing a fresh, light-pink wine from Provence, the Loire Valley, Spain or Portugal. These uncomplicated styles look lovely in the September sun and will adapt to any cuisine, from Chinese noodles with rich soy sauce to Lebanese roast aubergine and creamy hummus. Rosés are also remarkably successful when confronted with the wine lover's problem dish: salad with vinaigrette.

Visiting the Vineyards of Bordeaux

FRANCE

OPPOSITE, TOP LEFT: The village of Cadillac was a source of excellent dessert wine, but as fashions change it now produces more dry white table wine.

OPPOSITE, BOTTOM LEFT: The châteaux of Bordeaux are usually sturdily built and functional, rather than fairytale creations.

The map above shows the French *département* (administrative district) of the Gironde. The highlighted area shows the Bordeaux wine-production area, home to more than 8000 producers.

Bordeaux can be bewildering. This is arguably the world's most famous wine region, yet it is not the easiest to navigate. With its 120,000 hectares (297,000 acres) of land under vine, yielding almost 6 million hectolitres (about 132,000,000 gallons) of wine, this is a region that has no fewer than 57 different Appellation Contrôlée zones. You will find magnificent first-growth reds, such as Château Latour; delicate dry white wines; a few pink bottles; and a stunning repertoire of 'stickies' – dessert wines that include such unforgettable bottles as Château d'Yquem.

From the scenic point of view, start your voyage in Bordeaux itself. Not long ago, this austere, Neoclassical city seemed devoted to commerce. The riverfront *quais* were forbidding to all except those in the wine trade. Now that has changed, and there has been a fantastic transformation of the old buildings and docks. A stylish tourist destination has emerged from the old industrial landscape, and it is well worth spending a little time

exploring here first. Apart from great open spaces, fine vistas and impressive buildings, there are some excellent restaurants and also *œnothèques*, wine shops where a wide range of local wines may be tasted. (For more information on the city and the region in general, take a look at bordeaux.com.)

Venturing beyond the city, you have the choice of the Left Bank of the Gironde river, the classic regions of the Médoc, Pessac-Léognan and Graves, with huge estates and famous, expensive châteaux. Or you can opt for the Right Bank, arguably more scenic, and home to some of the most successful wines of the past twenty years in the exclusive enclaves of Pomerol and St-Émilion. The well-known American critic Robert Parker has helped to champion Right Bank wines, starting with such names as Le Pin and Valandraud. In contrast to the Left Bank, estates here are quite small, many only 5 hectares (12 acres), compared with a typical 50 hectares (123 acres) across the water.

Top, above and right: The Sauternes area has lovely cycling routes.

You will never be far from water in the Bordeaux region, be it the Dordogne or Garonne rivers or the Atlantic Ocean. Visit Arcachon and the immense Dune de Pilat, and marvel.

Overleaf: The river mist rising towards these sweet-wine vineyards in Bordeaux encourages the growth of 'noble rot' (*Botrytis cinerea*), which means the grapes are suitable for the making of such dessert wine as Sauternes.

The best option is to take a look at both sides, bearing in mind that a journey to the tip of the Médoc can take quite some time. If you plan to drive up through the famous villages of the Médoc, it is worth making some appointments before you go if a visit to a particular property is desired. There is rarely any wine to taste at the very finest vineyards, as the wines are presold via dealers in Bordeaux called *négociants*. Lesser-known names will advertise tasting opportunities, and it is good to make comparisons. From the city out to the village of Margaux the scenery is not thrilling, but then make sure to keep an eye open for the lovely châteaux of Palmer and Margaux itself – a first-growth claret, according to the important Classification of 1855, when the fine wines of the Médoc were classified in five categories of quality, with first growth being the top wines. At that time Ch. Latour, Ch. Lafite Rothschild, Ch. Haut-Brion and Ch. Margaux gained first-growth status; Ch. Mouton Rothschild joined them in 1973.

A little farther north is St-Julien, where there are three Léovilles: Barton, Las-Cases and Poyferré, all of them worthy of note. Ch. Léoville-Barton is owned by the urbane Anthony Barton, of Irish descent and the epitome of the Bordeaux grandee, evincing a blend of French elegance and British public-school charm, despite the fact that his family (and many others in the region) first settled here hundreds of years ago. Continue on to Pauillac, which has no fewer than three out of the five first-growth châteaux: Latour, Lafite Rothschild and Mouton Rothschild. The wines here tend to be austere in their youth, refined and long-lived, with the typical finesse of the Cabernet Sauvignon grape, which predominates in this region.

From here, backtrack to the Right Bank, by taking the ferry across to Blaye. There is an agreeable drive down the banks of the Dordogne through Bourg until you reach scenic Fronsac. There are affordable red wines galore to be tasted in these regions. Then travel onwards, through Libourne, a historic centre for shipment of wine

upriver back in the days when exports were by barrel rather than estate-bottled case. Immediately north of Libourne is the tiny settlement of Pomerol, where land is so sought after that only churches and schools still have recreational space. Every other inch is colonized by vines, mostly Merlot. The wines of this sacred enclave are hugely popular and very valuable. They have fruit, depth and an immediacy of charm, even when young, which is very attractive; plus the ability to age superbly.

Be sure to spend some time in the beautiful village of St-Émilion, a photographer's delight. The charming stone buildings are perched on a hill, and it is sensible to wander around on foot. Ch. Ausone is one of the top names and the topmost vineyards, while the other famous cellar is Ch. Cheval Blanc, down on the plain below. Around the town are 'satellite' villages with many excellent, affordable wines and tasting opportunities.

If time allows, Castillon and Bergerac (the latter just outside the official Bordeaux region) are also attractive market towns with good wine estates all round. Then there is the expanse of the Entre-Deux-Mers zone – between the 'two seas' of the Dordogne and Garonne rivers. Here you will find some terrific dry white wines made with Sauvignon Blanc and Semillon, plus attractive sweeter styles. But for the real classic dessert wines, cross the Garonne and visit Sauternes and Barsac, a pretty, hilly region with amazing, rich, sweet white wines made from grapes affected by the 'noble rot', *Botrytis cinerea*, which has the effect of concentrating the flavours of the grape and adding a faint whiff of wild mushroom aroma on the nose. Conclude your tour with a look at the Graves area and then Pessac-Léognan, where you will find the fifth first growth: Ch. Haut-Brion, right on the doorstep of the city of Bordeaux. And, if you fancy some pampering after your journey, stop at Ch. Smith Haut Lafitte, where Les Sources de Caudalie features grape spa products and fine dining.

Fairs and Festivals

· ·

September in the northern hemisphere is a month when wine festivals proliferate. Visit the website of virtually any region to find news of how the locals plan to celebrate the vintage.

In France, try the two-day Jurade de St Émilion, when cellar doors and estates are opened to visitors for tastings and other festivities: vins-saint-emilion.com

For a more unusual vinous destination, try Budapest, where the fine wines of Hungary are enjoying a renaissance. Be sure to sample the famous sweet Tokay: budapestadventure.com

In Ontario the weather is delightful and the grape harvest is not yet over: the famed Icewine grapes are kept on the vine until they actually freeze. Try this fine dessert wine and other impressive Canadian bottles at the Niagara Wine Fair: niagarawinefestival.com

Bernard Magrez

a Bordeaux wine baron, France

· ·

Bernard Magrez runs several prestigious wine estates in Bordeaux, and also has interests in Spain, Chile, Argentina and California. His entrepreneurial approach to the wine world can attract controversy.

bernard-magrez.com

You own thirty-five vineyards and have a reputation as a 'Baron of Wine'. How did you achieve this?

I am fortunate enough to own some top *crus* in Bordeaux, and I realized that wine lovers enjoyed learning about these wines and sampling them. So I decided to diversify and move into Portugal, Spain, South America and the Napa Valley, until eventually I amassed a total of thirty-five properties.

How did you start in the wine business?

I started some forty years ago with my own small, simple wine business; then I was fortunate enough to acquire some fine Bordeaux châteaux. Since then I have diversified and moved into a range of aspirational wines.

Now you are starting in the wine-tourism business. Can you tell me about your approach to this?

I noticed that this is a great way to introduce our wines to clients. In Spain there were some excellent examples of wine tourism. I then decided to go for this concept in Bordeaux, but I chose to concentrate on a high-end proposition, to reflect the quality of our châteaux, including Fombrauge and other top

names. It was a conscious decision to go for the top of the market.

From your entire collection, which wine would you choose to drink at your table?

I answer your question with two responses. One choice would be my wine Château Pape Clément, the other a wine from a region near Barcelona/Tarragona, called Priorat. Wines are made there on a very difficult, sloping *terroir*, very dry too. These are marvellous red wines. A winemaker is an artisan, always looking for excellence in his work. When he achieves this he becomes an artist, but wine is always a cultural product, not just a luxury commodity; it is a product of the land and of effort.

If you were a visitor to Bordeaux with some money to spend, what would your wine tours have to offer that is extra special?

Our 'mission' with the wine tours is to offer the maximum of pleasure, involvement with the history of the château, and then the possibility of dining and staying in the château itself. Visitors can travel in a vintage classic car to have a picnic, and spend more time

relaxing in this lovely environment. There is also the opportunity to take helicopter tours, a tasting lesson, or even a blending lesson, making your own 'assemblage' of wines and comparing these with the actual wines from the château. This gives people a real chance to understand wine much more profoundly.

Are you giving away trade secrets? This is not how Bordeaux usually operates?

These days, men and women are changing, their aspirations are quite new and different. What we are looking for now is accessibility, more information. For instance, my name appears on all my wine labels: this is not just vanity, but a desire to show the client that the wine can be trusted. Seeing the name of the owner on the label is very useful. I've been doing this for six years, due to public demand. People have always wanted to know how to tell if a wine is of good quality, and what could be better than to show the actual man who is responsible for its production? This inspires extra confidence.

Should wine and food be inseparable?

People talk a lot about the aromas of wine and their affinity with food, especially the great cuisine of France. Our mission should be to offer ideas on how to pair our fine wines with some of these great dishes.

You are very willing to speak openly, unlike many in Bordeaux. Would you see yourself as an international ambassador for Bordeaux?

There is such a range of properties in Bordeaux; each family has such different *terroir* and its own style of wine. For us, with our range of properties, we concentrate on pleasing the wine consumer, and we seem to be doing something right — so the figures tell us!

October

Vintages and Vats

....

A winegrower in upper New York State lovingly cradles a bunch of Cabernet Franc grapes.

This is the area of wine production that is most mysterious to the wine amateur. Talk of vintages and investments, wood ageing and oxidation, all sounds very challenging, and there are many restaurant sommeliers who seem to like nothing better than to intimidate their clients with the 'magic of vintage'. Yet there is no need to feel daunted. First of all, bear in mind that most of the wines that are being fermented, racked and stored in the month of October (at least in the northern hemisphere) will not be vintage-dated at all; many will be blended; and only a very small minority will be kept, reverently, in cellars either at the winery or in private homes. Despite those snooty sommeliers, the era of a wine being only as good as its advanced age is over. Most wines that we enjoy at home have been bottled and sold for almost immediate consumption. Many now have screwcaps to ensure absolute freshness and no hint of cork taint. This does not mean that the wine with a screwcap is inferior; it is simply at its peak of drinkability, and does not need to be stored for any length of time. The effort that goes into achieving this may be just as great as the work involved in creating a wine that has been aged for years in oak barrels.

The lovely fruit and freshness of a good white wine are all too easily ruined in the winery. In creating the great majority of whites and most pink wines, the aim is to retain the essence of the grape. This is achieved by judicious use of modern technology, which enables controlled-temperature fermentation, careful fining and filtering, and advanced bottling techniques. There is a fine line between a 'clean', fresh and flavourful white wine and a 'neutral' one, where all that glorious grapiness has vanished owing to overmanipulation by the winemaker and too much sulphur has been added to avoid spoilage. Try an inexpensive Petit Chablis against an example from a top shipper, such as Laroche or Durup, to understand the difference.

A fine, vintage-dated white wine is now therefore relatively unusual, and a wood-aged white even more so. Some of the finest oak-aged white wines in the world come from Burgundy and are produced from 100 per cent Chardonnay. The subtlety of the French oak barrels is just rich enough in added flavour to mean that the wine will preserve and improve its character for up to ten years (or even more for the finest wines). What is important is

Storing wine bottles in an orderly arrangement avoids a good deal of frustration later on; this cool tiled floor is ideal in helping to maintain correct temperature in the cellar.

selection: the winemaker will select only those individual vats of wine that show potential for this type of treatment. In some years there will not be sufficient fine new wine to merit the effort, which is why vintage dating does have relevance. It is a costly process to store wine, pay taxes on it and employ expert tasters to check its progress; that aged Puligny-Montrachet on a wine list is really worth the expenditure.

One of the regions that illustrates the 'vintage and vat' principle very clearly is Champagne (see 'January', pp. 20–25), where both red and white grapes are used to produce a blended wine that is then refermented and aged before sale. The level of experience needed to judge a youthful wine and determine that it will be good enough to appear on shelves as a vintage is very extensive – and not every year is a vintage one in this part of France.

If one scales down, and looks at a vintage date on a red or white wine from any part of the world, it is sometimes useful to know if that was a 'good year', but really only for marginal areas where the climate is tricky. These include Champagne, much of Germany, Chablis, northern Italy and cool-climate areas in California, the Pacific Northwest, New York State and New Zealand. The great majority of wines we buy are from regions where the weather conditions are gloriously consistent: much of Australia, Chile, Argentina, California (apart from the aforementioned cooler areas), Italy, Spain and France. Instead of considering a particular year, think of what you are looking for in a wine. Whites need that freshness of youth, so be wary of any bottle older than five years, unless it is a dessert wine or a notably fine label.

Red wines carry a reputation for improving year on year in bottle, but this is a dangerous assumption. Unless

LEFT: Fine vintage sparkling wine is produced all around the world, not just in Champagne; this is a Riesling Sekt from Germany.

RIGHT: Champagne ages better in magnums than single bottles. The sparkling wine is usually decanted from these larger bottles into small 'splits' for single servings or into half bottles.

OVERLEAF: In these Nuits-St-Georges cellars, fine Burgundies age in barrel and bottle.

the wine was made to last, possibly aged in barrel, it will deteriorate rapidly once it reaches its peak. Most everyday reds, like white wines, are at their best within about five years of bottling. You will generally find, however, that red wines from the cool climate regions mentioned above may benefit from keeping longer, if they come from a good producer, as they can take more time to show their full complexity. Bordeaux is a notable example: the red wines are drinkable young, but even the most basic Bordeaux Supérieur can take on more interesting depth after several years of being aged in bottle.

Grape variety is another factor to consider when thinking about ageing wines. Chardonnay from certain regions, such as Burgundy, will keep for many years; Chenin Blanc has astonishing longevity in the wines of the Loire; but Muscadet, Vinho Verde and other light whites rarely benefit from storage. They start to deepen in colour and lose the crisp acidity that lends them charm. Certain red wines are also intended to be youthful and fresh, and most varietals today are made with fruit in mind and designed for easy drinking. Wood-aged Merlot and Cabernet Sauvignon from Bordeaux contrast strongly with examples produced in the Veneto or Australia's Barossa Valley, which are absolutely designed for early drinking.

Many Italian red wines, however, are made with local grape varieties that taste slightly bitter when young, so allowing them to age suits some palates better. The Anglo-Saxon wine drinker has traditionally liked older wines; French wine producers, preferring wines that are sharper and have more 'bite', generally smile at our tendency to seek out older vintages. Overall, the trend is away from that long wood-ageing and the complex, fragrant bouquet that develops as a result. An aged wine feels lighter on the palate, and in this era of high-alcohol, fruit-driven wine, it is a challenge to taste and appreciate. Yet it is worth cultivating the ability to 'understand' a fine, delicate red Burgundy or a rich, mineral-quality Riesling from the Rheingau that have been aged some years in bottle. These are known in the wine world as 'meditation' wines, deserving of time and effort to appreciate. Take some time out from your routine and sip slowly and reverently. That work by the winemaker and care from the cellar master result in an experience that truly engages the senses.

Matching Wine with Game

Roast partridge on a potato cake, and perhaps served with cabbage, is the ideal dish to accompany a fine red Burgundy.

Whether you are the intrepid type who hunts down dinner yourself, or you are content to let someone else do the hard work, there are all kinds of delicious dishes based on game that work very well with wine. Thanks to the fact that we all seem to be seeking lower-fat meats, there is ever-increasing interest in swapping, say, pork for venison or farmyard duck for its leaner wild version. There is also an emphasis on eating food that is truly local, and game fits this bill admirably.

Look to the wine regions for inspiration about how to cook, serve and match game dishes with suitable bottles. There is a huge range of game out there, and it need not be costly. In France and many other European countries, the humble wood pigeon (as opposed to the tough feral pigeon found in town squares) is celebrated as a succulent delicacy. Similarly, hare and rabbit are both prized as a base for elaborate sauces or made into complex pâtés or terrines.

The passion for eating tiny songbirds seems to be waning in Europe, much to the delight of those who appreciate these creatures in their gardens at home; but quail, partridge and pheasant are still generally raised for the pot, and all pair very well with full-flavoured wines. Then there is wild duck, and the rarer birds, such as snipe, grouse and woodcock, all of which have their devotees.

The wild boar is a phenomenal pest in a vineyard: many of the finest Chianti estates dread the onslaught of these beasts, which can dig up newly planted vines overnight. In revenge, there are many recipes for wild boar in the region, with *cinghiale* featuring large in delicatessen and restaurant alike. Visit Siena to see a dramatic advertisement for the delights of boar sausage — a couple of fierce boar heads appear above one shop door. Wild boar range across a good many other wine regions, and appear on menus in Germany, Austria and France. Curiously, their rich meat tastes just as good with dry white wines, such as Riesling, or Austria's Grüner Veltliner, as with the more obvious reds.

Then there is venison: deer are frequently bred on the farm now, but most enthusiasts declare the wild version to taste better. Whichever you try, this lean meat is a wonderful foil for all types of red wine and, owing to its lack of grease on the palate, does not need as much

acidity to counterbalance it as many other meats. You can therefore indulge with chocolatey reds from the Douro Valley, Australia or California.

In the British Isles, there is a long tradition of game cookery, and the British excel at making game pie, serving partridge with cabbage, or presenting grouse with all its eccentric side 'trimmings', such as bread sauce and game chips. As a general rule, a completely simple roast-game dish tastes best with such classic wines as red or white Burgundy, claret (red Bordeaux) or fine Tuscan reds. When it comes to more complex pies and terrines, there is an opportunity for more adventurous wine choices, ranging from New Zealand Pinot Noir to New York Cabernet Franc – via Lebanon and Virginia.

Below are some wine-matching suggestions for game dishes.

Food	Wine
Grouse	Serve with finest claret from Bordeaux or full Spanish reds, such as Priorat.
Hare and rabbit	Serve with Italian Barolo, Barbaresco or any wine made with the Barbera or Nebbiolo grape worldwide. Or try a rich Rioja Reserva.
Pheasant and partridge	Delicious with good Pinot Noir from any region.
Venison	Goes well with fine wines from the Douro Valley of Portugal, South African Pinotage or Australian Shiraz.
Wild duck	A perfect match for fine red wines from the Rhône Valley and from Provence, including Bandol. Also goes well with top California and Australian Cabernet Sauvignon, not to mention fine Bordeaux reds, such as those from the Médoc.
Wild pigeon	Serve with a good Merlot from any region, or a rich rosé.

Finally, if you are serving leftover game as 'cold cuts', the recommendation from famed wine writer Hugh Johnson is to sip some mature vintage Champagne. And who are we to argue with him?

Visiting the Vineyards of New York State

The map, left, shows the American state of New York. The highlighted areas show the main wine-production regions covered in this feature.

As the leaves start to turn colour and take on their autumnal hues, the vineyards of New York State are at their most attractive. The harvest has been safely gathered in (apart from some late-harvest wines, including an occasional American Icewine, made with grapes frozen on the vine). It is a perennial source of amazement to wine lovers that there are such diverse and intriguing wines available within the state of New York. One reason why this is so startling is simply that most of the wines are sold either direct from the winery tasting room via wine clubs or to local restaurants and hotels. This leaves precious little to spread around, which is why this region is ideal for wine tourism. It is also a frontier country for winemakers, a place of experimentation, and therefore an interesting destination for the wine visitor.

Unlike a high proportion of their peers in other parts of the wine world, New York winemakers do not generally have to travel the world, marketing their wares. Instead, many can be found in the tasting room, chatting to consumers and explaining the diverse range of wines on offer. Let us begin our travels in Geneva, at the heart of the Finger Lakes region. Here on Seneca Lake there are wineries galore, for the most part established during the last thirty years, and all warmly welcoming to visitors.

Grapes have been grown in the region for very much longer: there are native varieties, *Vitis labrusca*, that helped to sustain life for the Iroquois who once lived here and that are still cultivated today. Today the Concord and Niagara varieties still thrive and are sold as modern wines, but the exciting story here is the coming of *Vitis vinifera*. This is the species that includes all those familiar grapes, from Chardonnay to Zinfandel, and is now yielding some excellent wines here and in other parts of the state.

The generally acknowledged pioneer of *vinifera* in New York State was Ukrainian-born Dr Konstantin Frank (1897–1985), and his winery is still a market leader today, true to his original principles. He was viewed as something of an eccentric when he arrived in the United States in the 1950s, speaking no English but determined to make his way in the wine industry and to use his scientific wine knowledge. The Frank winery was established at Keuka Lake, another of the 'fingers'. (These are said to be those of a Native American god who

BELOW, TOP LEFT: This view of the King Ferry Winery near Cayuga Lake shows how vines are just one part of the agricultural landscape of upper New York State.

BELOW, BOTTOM LEFT: Belhurst Castle typifies the aspirations of wealthy nineteenth-century residents around the Finger Lakes; today it is a romantic hotel with its own winery.

BELOW: This New York State winery gazebo was built with wine tourism in mind; this is a vital part of the economy here, as most wine produced is sold direct to businesses and consumers within the state.

OPPOSITE, CLOCKWISE FROM LEFT: The sun highlights the attractive greenish glints of healthy Riesling grape bunches.

Recently planted vines at Anthony Road Wine Company on Seneca Lake demonstrate that the region is constantly evolving in response to consumer demand.

Chateau Frank was where the Finger Lakes winemaking story began, in the 1960s, and it is still a powerful influence today.

Pinot Noir is a fussy grape that can do well in the cool climate of the Finger Lakes.

left his imprint on the earth in this way.) The grape type most usually associated with Dr Frank is Riesling, and these wines are still some of the most impressive at this winery and many others in the region.

Travelling around the Finger Lakes area is easy, and there are established wine trails to guide the visitor. Take a look at newyorkwines.org for guidance before setting out. Some of the wineries on Seneca Lake that should not be missed include Fox Run Vineyards (see Interview, p. 164), which makes excellent Riesling, Cabernet Franc and even port with its own 'hot room', Australian-style; Anthony Road Wine Company, which has a German winemaker and features superb Riesling, Gewürztraminer and a Finger Lakes version of sweet *Trockenbeerenauslese* that is outstanding. Continue on through the town of Seneca Falls, said to be the inspiration for the film *It's A Wonderful Life* (1946), on to Cayuga Lake and the agreeable King Ferry Winery, where dynamic Pete Saltonstall and his young female

winemaker are experimenting with wines that range from delicate dry rosé made with Cabernet Franc to rich, sweet dessert wine produced from hybrid grape Vignoles.

Drive on via Sheldrake Point and try its excellent Pinot Gris and Gewürztraminer; then take time to sample the classics from Dr Konstantin Frank. This winery takes time to find, and in consequence there is no charge for tasting. Start with the unusual Rkatsateli, a grape originally from Georgia (in the South Caucasus), the oldest wine region in the world. Here it tastes modern, faintly fragrant and refreshing. Then linger over some excellent Rieslings, some made bone dry, some with a hint of sweetness; the latter style works perfectly with food, especially spicy Asian cuisine. Another source of superb Riesling in the region is Hermann J. Wiemer. All the wines from the area are modest in alcohol, thanks to the tough climatic conditions, and complement great dishes rather than warring with them, as some higher alcohol wines from other regions can do.

Long Island

Not simply the location for F. Scott Fitzgerald's *The Great Gatsby* (1925) and modern-day beach opulence, Long Island is a serious centre for winemaking, with a great deal of passion (and cash) being expended on making the very finest varietal wines the East Coast has to offer. Knowing that there is a moneyed clientele within easy reach certainly helps: in season, the tasting rooms on the North and South Forks of the Island are packed. Do not be daunted by the choice among all these fascinating wineries; it is an easy area to tour around, and it is well worth exploring both sides of the island.

From early plantings thirty years ago by the Hargrave family (now the Castello di Borghese winery), followed by Bedell (another premier winery) and other pioneers, this spit of land has aspired to take on a role as America's Médoc. It has sandy, loamy soils that drain well and are ideal for grapes. Where once potatoes and cauliflowers grew, now there are vines of all varieties. Despite the comparison with Bordeaux, this is a strong region for white wines, and many top winemakers are experimenting with such unusual grapes as Tocai Friulano (at Channing Daughters on the South Fork), Muscat and Viognier, as well as cultivating the more familiar Sauvignon Blanc and Chardonnay. Channing Daughters is a modern winery with a simple tasting room where you can sample some remarkable and very imaginative wines.

Strike an immediate contrast in visual style by visiting Wölffer Estate Vineyard in Bridgehampton, a glamorous villa with spectacular views and some outstanding wines made by Roman Roth, including excellent whites (just to counter that rumour about

Above: As there are no historic traditions of growing grapes on Long Island, such local winegrowers as Macari happily experiment with a wide range of grape varietals.

Left: The Wölffer Estate Vineyard produces a fine Merlot; winemaker Roman Roth has exacting production standards.

white wines not hitting the quality mark on Long Island) and some deep, flavourful Merlot made for keeping.

Next, travel over to the North Fork, where the choice of wineries is huge. Try contrasting a classic, established set of wines at Bedell or Pindar with somewhere like Macari, where the owners are passionate about their land and use the most responsible farming methods possible, including keeping their own cattle to ensure a good supply of manure for the vines (see Interview, p. 165). The result is honest, full-flavoured wine with distinctive character.

These regions are coming of age, and no longer trying to imitate anything made in Europe, concentrating instead on making original wines that will sell direct to the public and also appeal to the discerning restaurant-goer. Visit, and relish their enthusiasm and the region's beauty.

LEFT: Investment in quality at the Wölffer Estate Vineyard has paid off in the wines; the tasting room is a stylish place to visit and appreciate them.

BELOW: Long Island is a source of excellent produce, often sold on wayside farm stands; this classic barn at Macari Vineyards echoes this tradition.

Interview

Scott Osborn

President of Fox Run Vineyards in upper New York State, USA

....

Scott Osborn has worked in the wine industry since 1980. He purchased Fox Run Vineyards in 1993.

foxrunvineyards.com

Please give me a potted history of Fox Run Vineyards.
When I purchased Fox Run in 1993, it was a 10-hectare (25-acre), young vineyard, and that first year we made about 10,456 litres (2,300 gallons) of wine, or 1000 cases. Since that time, we have grown to 20 hectares (50 acres), and we bottled 20,000 cases of wine this year. Our visitor count has gone from 7500 people in 1993 to 80,000 people tasting wine now. With the help of John Kaiser, our vineyard manager, and Peter Bell, our winemaker, we have built a quality reputation throughout the world. We are also leaders in vineyard sustainability, and leading advocates of the 'Buy Local Wine' movement.

What factors influenced your choice of grape varietals?
This is a cool-climate region, and certain varieties grow better than others. Back in the early 1990s, we knew generally what would survive, but not necessarily what would be great. Now, after all these years, based on how the vines have survived over the years and how consistent the wine quality has been from year to year, we know what varieties are best suited to our climate and soils, and have narrowed it down to about five varieties: Riesling (the king), Chardonnay, Lemberger, Cabernet Franc and Pinot Noir.

What would make a wine tourist divert to visit you?
We are a favoured destination because we give great tours of the vineyards and winery. Our staff are well trained and actually experience many of the different aspects of what they talk about.

Joe Macari

of Macari Vineyards, Long Island, New York State, USA

The Macari family owns the waterfront estate on which Macari Vineyards is established. The vineyards were first planted there in 1995.

macariwines.com

Tell me the story of Macari Vineyards.
The property that Macari lies on has been in the family for well over forty years. Joe Macari Sr had always wanted a vineyard, and we have his actual vineyard site plans from the late 1970s. Fast forward to 1994, Alexandra [Joe Macari Jr's wife] and I and two children move out from New York City and begin clearing the land in preparation for 1995 planting. This planting was 27 hectares (67 acres) – unheard of on the North Fork. They thought we were crazy and that we were thinking 'too big'. The next year we planted 16 hectares (40 acres) ... not too bad for two city guys.

How do you approach vineyard management?
Day by day, is the simple answer. We are gentle with the use of sprays; we'd far rather be spreading compost. We do the

sprays that are required. I give the soil what it needs and the plants benefit from the nutrients in the soil. Feed the soil instead of feeding the plants, and the plants are able to take what they need from the soil.

How did you select your grape varieties?
We chose mostly French *vinifera*, primarily Merlot and Chardonnay; we had acquired 4 hectares (10 acres) of twelve-year-old Merlot with our winery building in the Mattituck Hills. From then we planted several other varieties. Cabernet Franc and Sauvignon Blanc were thought to do well with a maritime climate, and turned out to be excellent. A lot of trial and error and hard work – and there are probably more grapes that will do well here.

Could you briefly summarize your winemaking philosophy?
We want to offer wine of the vineyard, or *terroir*, not of the winemaker. We try to make wines with little manipulation, so that they end up exhibiting fresh-fruit characteristics, not masked by oak.

Where do you see Macari in ten years' time?
I'd envisage the vineyards getting fine-tuned, more volume and better quality as we learn more. Increasing volume up to 20,000–25,000 cases, and keeping true to our vineyard standards.

November

Storing and Serving Wine

OPPOSITE: Ancient bottles of Burgundy rest in the cellars of the Domaine du Clos Frantin in Nuits-St-Georges, famed for its red wines made with Pinot Noir.

RIGHT: A well-stocked home cellar deserves to be catalogued in a cellar book, or its modern-day computerized version.

While the work of protecting the vines from late frosts is being undertaken in the southern hemisphere, in the northern, after all the energy and effort of fermentation, November is the time for wines to be transferred to the cellar or *chai* (above-ground storage) to rest and mature before they are ready to bottle. It is an anxious time, as this is when the winemaker has to ensure that the wines stay stable and healthy throughout the winter. This is far easier to achieve with modern technology (see 'September', pp. 133–37), but not long ago the natural frosts of the colder months of the year were essential to ensure that wines could 'run clear', dropping such vestiges of fermentation as dead yeast cells.

Storing

At home, indoor entertaining is a way to enliven dull, cool November days and evenings. There is a great deal of debate about how best to store and serve wine, and the general principles are really fairly simple. As wine in the bottle is still 'alive' and able to undergo change, your bottles need to be kept somewhere that is dark, constant in temperature and not subject to sudden vibration. Bright light is the enemy of wine, as is high temperature (or freezing cold); sudden movement also stirs up the bottle and can mean the wine starts to referment, with disastrous consequences. In a modern home, unless you have the funds to invest in a purpose-built cellar, it may be easiest to store all your bottles in a special wine fridge, and bring out the reds well in advance of serving, to allow them to warm up a little. The average centrally heated home has its thermostat set at 20°C (68°F) or more, which is at the upper end of a comfortable storage temperature for wine. The highest you should go is about 18°C (64°F), if your wines are stored in a spare room; and try to ensure the room remains at a constant temperature all year round. If the wine warms up suddenly, it may start to oxidize and then spoil.

If you intend to keep wines for a while, it is fun and useful to create a cellar book. Assuming your wines are in a rack, number each slot and then use graph paper (the squares equate to wine racks) to pencil in the name of each wine, with a code to indicate when and where you bought it: for example, 'Nuits-St-Georges 2007, BBR 10/10' would be a red Burgundy acquired from well-known London wine merchants Berry Bros. & Rudd in October 2010. Of course, if you are not adamant about literally keeping a record in the cellar, then a table created on your computer will do just as well.

It is always interesting to sample a bottle from a case of wine fairly early in its life, and then test again later on. In this way, you can decide if you are really enjoying your expensive investment, or whether you might want to sell the remainder at auction. This would apply only to really high-quality 'investment' wine, such as Bordeaux, Burgundy, or rare California or Australian wines. At the opposite end of the spectrum, if you buy at a bargain price some full, tannic red wine that shows promise, it is fun to rebottle some into halves, as it will then age more rapidly and may taste smoother in just a matter of months.

LEFT: The willow hoops on this Burgundy cask help the cellar worker to move the barrels with greater ease.

OPPOSITE: A decanter need not be cut crystal; for fine wines, a clear glass style shows off the beauty of the colour.

Serving

It is all too easy to take away all the pleasure of a wine by serving it at the wrong time, in the wrong place or at the wrong temperature. All the clichés about wine tasting better on holiday tend towards truth, because if you bought the wine locally, you tasted it in its ideal context. To create the same conditions at home, plan ahead, and think about the wine you will be serving. Consider whether it will be necessary to decant the wine. If you have a very elderly vintage port or fine claret, decanting may help to separate the wine from its sediment. Traditionally, this task would be performed by a butler in a dark cellar, holding the bottle carefully over a candle to ensure no bits and pieces made their way to the decanter. Today, a simple funnel and some bright ambient light are useful. Real aficionados of fine wine have been known to strain the very last of the bottle through a clean coffee filter and save this portion for the host.

Back in the reality of modern-day life, these rituals are rarely necessary, unless you fancy decanting a simple screwcap bottle of wine into a carafe for ease of pouring. Temperature, however, is absolutely key. The domestic refrigerator generally hovers at around 6°C (43° F), which is a good serving temperature for sweeter white wines, such as Sauternes, and for sparkling wine. Top Champagne deserves to be served at up to 10°C (50°F) to showcase fully its toasty aromas; the same is true of all good-quality white and pink wines. A light red, such as Beaujolais or Valpolicella, can taste absolutely delicious when served just a fraction warmer, at around 11°C (51°F). Fruity, everyday red wines are most attractive and enjoyable when not too warm, at about 12°C (53°F).

Wood-aged Chardonnay can be treated almost as a red wine, and served at 12°C (53°F) or even slightly higher. Above this temperature come the warm, rich and classic red wines of Burgundy, Bordeaux, Italy and Spain. The very finest reds actually taste good when served right up to 18°C (64°F) – almost, but not quite, modern room temperature. Finally, do not be tempted to warm up your reds by the fire, the stove or the radiator. Unless the wine is very simple and rustic, this is a sure way to make a good bottle over-alcoholic and dull on the palate. Patience is the key; plan ahead, and allow your bottles of red to acclimatize gently to a warmer room.

Seasonal Celebrations

The short, dark days of November are a time when we all feel the need to raise our spirits. There are various traditional times for celebration, including, in England, Guy Fawkes Night, when fireworks are set off to commemorate a failed attempt to blow up the Houses of Parliament in 1605; in the United States, Thanksgiving celebrates the early Pilgrim Fathers' survival in their new land. Without probing into the politics, both occasions are great opportunities to eat, drink and get together with friends and family of all ages.

Mixed age groups require a range of drinks, and this could be a good time to think about a family version of the cocktail party. Everyday wines need not be treated with too much reverence, and a full-flavoured red or white wine may be mixed with fizzy mineral water to make a Spritzer, or blend Champagne with orange juice to create Buck's Fizz (called Mimosa in North America). If the weather is chilly, a warming cup of mulled wine is perfect. It needs to be made with care, and to be kept really hot, as a tepid version is simply bitter and unpleasant. Choose a red wine that has plenty of flavour and fruit, such as Tempranillo, Shiraz or Zinfandel. For

each bottle you pour into your pan, add a cinnamon stick, three cloves, a few pieces of dried orange peel (dry them in the oven after you have been cooking, and while the oven is still warm), and dark-brown sugar to taste. Heat gently, avoid boiling, then serve at once. Add a dash of cognac or other grape-based brandy if you like a glühwein kick to keep you warm. Younger members of the family can be offered something similar but non-alcoholic: mix fruit juices with spice and a little sweetener (blackcurrant is especially delicious, with honey).

Now for a choice of menu. If you are having a bonfire for Guy Fawkes Night, there are certain foods that are traditional, including baked jacket potatoes (possibly cooked in tinfoil in the fire itself), good old British 'banger' sausages, well grilled, and something sticky and treacly, either toffee apples or gingerbread, afterwards. But prior to this, how about some warming home-made soup? One that goes down well with all age groups is made with tinned beans, including cannellini, butter beans and borlotti; simply purée them with sliced onion softened in butter, and add a sprinkle of dried or fresh sage, salt and pepper; thin the purée with warm

RIGHT: Pink wines need not be kept just for summer drinking; this cheerful California rosé will brighten up your seasonal table.

BELOW, LEFT: The red wines made with the Montepulciano grape in the Abruzzo region of Italy are reliable, well-structured reds. Be careful not to confuse them with the finer wines of Tuscany, the Vino Nobile di Montepulciano, made in the village of that name.

BELOW, RIGHT: Warm, welcoming red Rioja can charm even those who generally avoid red wine.

stock made with a quality cube or powder, such as Marigold. A small bowl or mug of this is certain to keep the hands warm.

Wines that go well with this sort of fare include hearty reds from Italy, such as Primitivo, Nero d'Avola or Montepulciano d'Abruzzo; punchy French reds made with Syrah and Grenache from the Languedoc or the Rhône Valley; Spanish full-bodied reds from Catalonia or Rioja; and a wide repertoire of New World reds, including South African Pinotage, Australian Shiraz and South American Malbec. If your guests prefer something lighter, choose a white wine that has plenty of roundness on the palate, such as California Chardonnay, or a Viognier from the South of France.

A Thanksgiving dinner is generally an indoor affair and can vary from informal, almost 'pot luck' style, with each guest bringing a dish, to quite stylish, with the table dressed in its finest napery. The actual menu, again, tends to the traditional, including roast turkey with all sorts of trimmings, and the famed pumpkin pie. Many of the classic side dishes are almost as sweet as the dessert, so the wines need to be chosen with care. As

your guests may well be staying the night, this is a moment to indulge, relax and enjoy the family gathering.

Start the evening with something crisp and light, to stimulate the palate without overburdening it before the feast. Champagne or fine-quality sparkling wine is one obvious choice, but another could be a glass of very good Chablis or Beaujolais, served with a few delectable cheese *gougères*. These are a savoury equivalent of profiteroles, and a feature of the Burgundy wine region; they are made by adding good-quality hard cheese, such as Gruyère, Cantal or Emmental, to choux paste.

This could be a moment to start the meal American style, with a colourful array of salad, including red radicchio, orange peppers and tiny cherry tomatoes amid the green; add some quails' eggs and tiny chunks of pâté and ham with some sliced olives and fresh chives to make a *salade gourmande*. Continue with the Chablis and Beaujolais, and make sure your dressing is not too vinegary. Use hazelnut or walnut oil with discretion to add a French twist, and serve with crisp baguette bread.

The roast turkey and its many side dishes deserve a truly fine wine accompaniment, and since this is an American tradition, choose California Chardonnay or Cabernet Sauvignon – the best you can afford. Alternatively, a Cabernet Franc from New York State, or a Petit Verdot from Virginia, would fit the bill admirably.

Pumpkin pie is a rich, spicy indulgence, and deserves a wine made with the Muscat grape to show off its complexity of flavour. One ideal match is Muscat de Beaumes de Venise from France, or a Moscato di Pantelleria from Italy; Australia also makes terrific Orange Muscats. Another idea is to end the meal in festive mode, with a glass of sparkling Asti Spumante, also produced with Muscat and low enough in alcohol for younger members of the family to take a sip.

ABOVE: Champagne is always the right choice for any celebration, and Chablis, with its crisp, delicate flavours, adds great sophistication.

LEFT: Fleurie, a fruity, deliciously light Beaujolais, should be served slightly cool.

OPPOSITE: Relaxing with friends and a glass of wine is the ideal way to mark a seasonal festivity.

Visiting the Vineyards of Burgundy and Chablis

The map above shows the *département* (administrative district) of Yonne, in central France. The highlighted areas show the Burgundy and Chablis wine-production regions.

This region of France may not seem overtly focused on attracting tourists, yet it is absolutely accustomed to welcoming a stream of visitors, year round. For centuries, the wines of Burgundy, Chablis, Mâcon and Beaujolais have reached their public by the simplest of means – personal collection. Unlike the giant quayside warehouses belonging to the Bordeaux *négociants*, where wine destined for overseas was held in barrel, here in landlocked Burgundy, wines travelled overland. There is therefore a long history of wine lovers making the trip from Austria, Switzerland, Belgium or Germany to fill their carts or cars with crates of fine wine. It was the Romans who established the first vineyards and made plans for a 'Rome in the West' based around the picturesque but now obscure town of Autun; today, those vineyards still thrive, and there is a charmingly old-fashioned feel to the way business is done.

Some of the wines from these regions are among the most sought-after in the world (for instance, Clos de Vougeot, Meursault and Nuits-St-Georges), and at the annual Hospices de Beaune auction in November they

LEFT: The steeper the slope, the finer the wine, is the rule of thumb in Chablis; the Grand Cru vineyards are vertiginous.

BELOW: The sloping 'shoulders' of these bottles are typical of Burgundy; Bordeaux bottles are 'square shouldered'.

reach phenomenal prices. However, many others are surprisingly affordable, and a visit to the area will uncover all sorts of unexpected tasting delights. Autumn is cool and misty, but charming as a time to visit, with welcoming fireplaces in the inns of Beaune, Chablis or the nearby city of Lyons. The cuisine of this region is substantial and hearty, with an emphasis on meats, including the refreshing *jambon persillé* (ham terrine with fresh parsley); the rich stew known as *coq au vin*; and, of course, *bœuf à la bourguignonne*, cooked long and slow with plenty of local red wine. Snails, spicy sausage and eggs are cooked with wine as well; try *œufs en meurette*, a dish of poached eggs with rich onion-and-red-wine sauce. To conclude, such local cheeses as Chambertin (also the name of one of the region's finest wines) and Époisses are pungently flavoured and a perfect complement to the earthy, truffly aromas found in the Pinot Noir grape.

Old stone walls are a major feature of the Côte d'Or vineyards of Burgundy; there are even stone-enclosed graveyards within the vines for the winemaking families.

An elegant Burgundian *manoir* at Domaine du Pavillon in Pommard; the use of the word 'Château' is rare in the region.

The wines of the region are almost all based on just two grape varieties: Chardonnay and Pinot Noir. Chablis is 100 per cent Chardonnay, as are the whites of the Côte d'Or, Burgundy's heartland. All the red wines are 100 per cent Pinot Noir, apart from those mavericks in Beaujolais who focus on their beloved Gamay, a fruity, distinctive varietal that tastes a little like jelly beans. Leaving grapes aside, understanding the labelling on a Burgundy can be tricky. Learn the names of a few key shippers whom you can trust, and seek out their wines. Examples of reliable names or labels would include Bouchard Père et Fils, Drouhin, Duboeuf, Faiveley, Louis Jadot and Labouré-Roi. As the vineyards themselves are often tiny, these shippers play a key role in assembling and marketing wines. You will notice that many of the villages have double-barrelled names – Gevrey-Chambertin or Pernand-Vergelesses – as do many of the families cited on the labels. Napoleonic code applies to inheritance here, and when a *vigneron* dies, his estate is generally divided among all the children, creating havoc with continuity of production.

The attractive, modest little town of Chablis is just a short drive from Paris, or a rapid ride on the excellent TGV trains, and is a centre of excellence for fine, dry white wines. Its name has been taken in vain by vineyard regions all over the world, but by law, wines of this name may be produced only in this tiny, isolated little corner of France. The steep-sided Grand Cru vineyards are tended with loving care, and they provide a great opportunity for an energetic autumnal stroll to appreciate how difficult they are to work. The climate here is fairly harsh and unforgiving in winter, and spring frosts are commonplace, yet the glorious minerality of the wines is worth the effort. When in Chablis, visit some of the well-known producers, such as Brocard, Dauvissat, Durup, Laroche

RIGHT: Meursault is one of the finest of white Burgundies; it is often aged in wood and has a long life in bottle.

FAR RIGHT, TOP: Combining wine tasting with gentle exercise is a perfect way to appreciate the vineyards; many wine-tourism companies now offer hiking holidays.

FAR RIGHT, BOTTOM: The Grand Cru vineyards of Chablis are so precious that some owners even run heated cables along the steeply sloping rows of vines to warm them during periods of frost.

and Michel, but do not neglect to visit the excellent local wine co-operative, La Chablisienne, as well. The same advice applies to the south of the Côte d'Or, down in the Chalonnais, where fine Pouilly-Fuissé white wines are made, and to the pretty, hilly region of Beaujolais, where the Georges Duboeuf visitor centre is admittedly commercial but also home to some superb wines.

Now to the heart of the matter: the two major zones producing fine red and white Burgundy in the long, narrow Côte d'Or. Driving south towards Beaune, you are in the Côte de Nuits, famed mainly for red wines. After a look around the scenic sights of Beaune itself, you will be in the white-wine area called the Côte de Beaune. There are endless opportunities to stop and sample, as well as the chance to wander a little 'off piste' into the hills of the Hautes-Côtes, where less expensive but often delicious wines are made by small producers. The famed Clos de Vougeot is a popular tourist stop, but may be a little too regimented for some; try the Château de Meursault, a wonderful structure of eleventh-century origin, where tastings of a selection of the wines in the range made by Patriarche will provide an instant overview of Burgundian wine for the wine amateur.

Remember that the fine Chardonnay and Pinot Noir of Mâcon or the Côte Chalonnaise can offer tremendous value, and the villages here are pretty and unspoiled. Look for the wines of Mercurey, Givry and Rully as you gently tour this classic French landscape.

Fairs and Festivals

· ·

The famed Hospices de Beaune in Burgundy, France, holds its yearly auction on the third Sunday in November; the wines can be tasted in the Hospices' cellar before the sale: hospices-de-beaune.com

This is also the month for Beaujolais Nouveau and a host of similar fresh, lively wines of the current vintage, fruity and not intended for keeping. In Italy, try the young Muscadet or Vino Novello from Puglia. There is a light-hearted festival, the Vino Novello festival, to enjoy in Leverano: discoversalento.org

In northern Italy near Bolzano, snow could well be falling on the vineyards during the Merano Wine Festival: meranowinefestival.com

By now it may also be snowing in Slovenia, where on 11 November they celebrate the Festival St Martin to welcome the new wines: http://en.slovenskifestivalvin.si

LEFT: The annual Hospices de Beaune wine auction is an entertaining and prestigious event that also benefits a long-established charity in this region of Burgundy.

BELOW: The fresh, fruity flavour of Beaujolais Nouveau is cause for celebration and has inspired all manner of eccentric events.

Alberic Bichot

CEO of Albert Bichot wines, Burgundy winegrowers, France

The house of Albert Bichot was established in 1831 and remains family run. It favours an organic style of viticulture.

wine.gg/Albert-Bichot

Like so many Burgundy houses, yours is a family affair. Can you tell me something about your history?

These days the family businesses are rarer. There's a simple reason: inheritance laws in France mean that the estates have to be divided equally among all the heirs, so vineyards are reduced in size until they literally disappear! Our company was established in 1831, and after six generations we are still quite visible, thanks to the efforts of my grandfather and father to reacquire areas of vine that had been transferred under these laws. So today, we are still a recognizable family. Compare us with Domaine Leflaive [famed white-wine producers, best known for Montrachet]; they now have no fewer than sixty-eight separate owners. Here at Bichot we have a true passion for the Burgundy region, including Chablis to the north and the Chalonnais to the south. We don't want to move away from our roots, but to stay and improve within our boundaries here.

Could you characterize the particular features that make Burgundy unique?

This is a region where we have to suffer for our wine! We have cold winters, and

fierce heat in summer. The Pinot Noir grape is notoriously fussy and fickle too.

Historically, Burgundy has been a region riven by conflict, not a rich, settled area like Bordeaux, with its access to the sea. As a result, we merchants and growers have developed strong relationships with our neighbours in Belgium and Switzerland. It is still the case that wine lovers travel down to Burgundy and fill their cars with select wines, just as happened in centuries past, when wines would travel overland to private collections, restaurants and hotels. This is quite different from the Bordeaux system of *négociants*, who sell in great quantity and create commercial blends.

Tell me more about your regional grapes and wine styles. Do you have any favourite wines?

I think a Premier Cru Puligny-Montrachet is a true expression of the finest white Burgundy. It is pure and clean, with a hint of that minerality found in Chardonnay here. This quality is at its most marked in Chablis, of course. We are not great oak-barrel enthusiasts at Bichot; we prefer

freshness in our wines. The reds have cherry and cassis [blackcurrant] aromas in their youth. All the wines have excellent acidity balance. We have vineyards in many areas of the Côte d'Or, Chablis and to the south in the Chalonnais, but each one has its own team, with a remit to make wine that reflects the *terroir*. We are not trying to make our wines conform to a 'house style' and restrict creativity.

How about wine tourism? What would a visitor find when visiting Burgundy?

Burgundy is absolutely, authentically French: we have wonderful food, which pairs well with our wines. We are also open to the world, on a trade route from north to south, and we welcome streams of visitors every year. Our vineyards are 'on the way' to so many places, from ski resorts to beaches. So we have a visitor centre in Chablis [at the Domaine Long Depaquit], and we are happy to see tourists at our other sites across the region. Far from being the exclusive, expensive enclave of legend, Burgundy is one of the most welcoming places to come and taste and buy wine.

December

The Ultimate Wine Selection

A great deal of wine changes hands over the festive season, and giving a bottle or even a case as a present is a traditional way to show appreciation to everyone from employees to the person who delivers your mail, not to mention all those awkward family members. Much of this wine is chosen for us by wine merchants, supermarkets or mail-order companies, yet this moment of giving has the potential to be exciting and to stimulate a real interest in collecting wine.

Assembling a discerning assortment of good wine is profoundly satisfying, and, unlike so many collections, it is also a very social decision. The wines you acquire will be the mainstay of many a dinner, informal lunch or party to come. There are essentially two levels of wine acquisition: fine or classic wines that are truly the best in their category; and the wines you like to drink on an everyday basis. Looking at a few good restaurant wine

The iconic 'gum trees' (eucalyptus) of Australia mingle with other tree species at these lush vineyards in Western Australia's Margaret River region.

lists will help formulate your ideas on both counts. Note how the house wines are well chosen to be affordable yet interesting, perhaps reflecting the region of the cuisine on offer, or including reliable varietals that have a wide appeal.

The more expensive offerings your sommelier may be proposing are sure to include some Champagne, sparkling wines, classic Bordeaux, Burgundy and probably fine Italian wines, with a smattering of representatives from the rest of Europe and the New World. (If you are in the United States, the mix is likely to show far more native wines, from California, the Pacific Northwest and New York State.) Wherever your restaurant, there are recurring themes of what is considered to be essential to a good wine list, and this should be your starting point for a collection at home.

How many bottles constitute a collection, and how are you to keep them in good condition (see 'November', pp. 168–69)? Cases usually comprise twelve bottles, so why not go for five of these as a comfortable start – a total of sixty bottles. Half of this number can be white, sparkling or pink wines, and the remainder red, with one or two bottles of port included. Port is a divisive wine, with, on the one hand, strong supporters, and, on the other, those who refuse to sample it at all. Yet when the scores of the famed Robert Parker, wine critic extraordinaire, are reviewed, the top ten consistently highest-scoring wines are headed by Taylor's Port, with 97.8 out of a possible 100 points on average. The remainder of the top ten include fine Bordeaux, such as Ch. Ausone, Ch. Pavie and the more obscure Ch. Bellevue Mondotte (all from St-Émilion); Ch. Haut-Brion Blanc, the white version of the famed red wine; Ch. Pape Clément Blanc, from the estates of Bernard Magrez (see 'September', p. 149); Ch. d'Yquem, finest of all sweet

Sauternes; and Krug vintage Champagne. Apart from Taylor's, the sister company Fonseca's port is also in Parker's top ten. The legendary Ch. Latour makes it in only at number 10.

You could do worse than sample all of the above, just to see what all the fuss is about. For the ports, buy a bottle from a good recent vintage and also a simpler version, such as a Late Bottled Vintage (LBV); the styles are very different and worth comparing. Remember that vintage-dated port needs to be kept for at least ten years after its harvest before drinking, whereas LBV is ready when you are. Similarly, if you buy a bottle of the other top red wines, check with your merchant if it is for keeping or current drinking. As a general rule, a minimum of five years is needed to soften the edges of a Bordeaux red wine; longer for a Latour.

Move beyond these horizons and consider how to round out your mini-collection of ultimate wines. Go back to that theoretical restaurant list and buy a couple of bottles of top-quality red Burgundy, such as Chassagne-Montrachet, from a good shipper, as well as a pair of great white Burgundies, such as Meursault. Add a Grand Cru Chablis and the best Beaujolais you can find; Morgon and Moulin-à-Vent are villages that produce wines that keep well. Complete the picture with a bottle of Pouilly-Fuissé white, the top wine of the Chalonnais district.

You have one Champagne, add another – for example a pink, such as the impeccable Billecart-Salmon, maybe a vintage-dated bottle. To represent the Loire Valley, go for a fine Sancerre, made with Sauvignon Blanc, and also a mature Vouvray, a wine often overlooked today, but made with Chenin Blanc, which has an impressive ability to age well. Then add a Hermitage from Guigal, a producer just outside Parker's top ten, but universally admired. Try Guigal's white Condrieu,

Top row, left to right:
Krug is an iconic brand in the Champagne spectrum, and well worth the outlay.

A fine Prosecco is the ideal summer aperitif.

Macari and other New York State producers are now the source of some excellent, lower-alcohol red wines.

Bottom row, left to right:
A vintage port, such as this example from Graham's, is an essential for a complete cellar.

Château Latour is among the five first growths from Bordeaux, long established as being of the highest quality.

The stylish presentation of the so-called Super Tuscan red wines, such as those produced by Ornellaia, is echoed in their rich, complex flavours, from classic Cabernet Sauvignon and Merlot grapes.

too – a classic expression of the Viognier grape. Complete your Rhône trio with Châteauneuf-du-Pape. From Provence, buy one really fine pink wine and one bottle of red Bandol.

Finishing your tour of France, select a fine wine from Alsace; Pinot Gris or Riesling, according to taste, from a good shipper, such as Hugel. Then add a couple of German wines, both Rieslings from the Rheingau, one of which should be a Kabinett and one a full, rich Beerenauslese. Then add another Riesling from the Mosel to see how they contrast.

Now you have Italy and Spain to play with. One red from Piedmont – a great Barolo is a good choice; one fine Chianti; one Brunello di Montalcino; then add a great Pinot Grigio from the north of the country (look out for those produced in Friuli-Venezia Giulia). Add to this a bottle of Vin Santo and a top Muscat dessert wine, for example from Pantelleria.

Spain deserves to show off its fine reds, including a bottle or two of good Rioja, a bottle of Priorat, one from Ribera del Duero, and – to represent the sparkling-wine industry – the best Cava you can find.

As the final European example, buy a bottle of Tokay, acknowledged as one of the world's great dessert wines, from Hungary.

Crossing the Atlantic, add the finest wine you can source from New York State; try one made by Roman Roth. Over to Oregon, where you must add a Pinot Noir to your collection. Next, California: so much to choose from, but a classic Chardonnay and Cabernet Sauvignon from a top cellar, such as Beringer, Joseph Phelps or Stag's Leap, come first; then a Zinfandel from such a specialist as Ridge Vineyards to represent this most typical of California grape varieties; plus a Fumé Blanc, for example from Mondavi, inventor of this style.

Opposite, top left and right: Brunello di Montalcino, from the Sangiovese grape, is among the finest Tuscan red wines and Antinori is a top producer of all Tuscan wines.

Likewise, Prunotto is a noted producer of Barolo, the finest red wine of the Piedmont region.

Opposite, bottom left and right: Having some sparkling wine such as this Italian Pinot Grigio on hand means you are always ready to welcome good news or unexpected guests.

Red Rioja wines exist in Reserva (at least three years old) and Gran Reserva (at least five years old) styles, and both are usually excellent value for money.

Above right: Tokay (Tokaji in Hungarian) was a favoured wine of the Russian tsars; after a period of decline under Soviet rule, this dessert wine is firmly back on the collector's list.

Right, left to right: A light style of Pinot Noir from the Finger Lakes region of New York; rich, spicy Australian dessert Muscat, good with chocolate dishes; and Penfolds Grange, a collector's red wine from this famed producer.

Complete the picture with a top sparkling wine from such a cellar as Schramsberg, and a Quady port.

From the New World, you will need a New Zealand Sauvignon Blanc and a Pinot Noir, and Australian Shiraz and Cabernet Sauvignon. If your budget runs to Penfolds (see Interview, pp. 204–205) Grange Hermitage, it is worth the investment. Australian Semillon and Riesling should also be on your list. From South Africa, choose a bottle of Pinotage and one of Chenin Blanc.

Finally, over to South America, where you need to acquire a great Chilean Cabernet Sauvignon, a bottle of Sauvignon Blanc and an Argentine Malbec.

Complete your selection of what the wine world has to offer with a bottle of American or Canadian Icewine, just to startle the palate.

Making the Most of Your Fine Wines

Make the most of your wine collection by serving wines with matching regional cuisines. Here, rich choucroute will be balanced by the crisp acidity of Alsace wine.

Once you have had a chance to browse the ultimate wine selection in the preceding pages, you can think about how best to showcase the fine bottles you may acquire. Since this is the holiday season, there may be more time than usual to plan and prepare interesting meals, and to ensure those special wines are shown off to advantage. Here are some suggestions as to how best to combine the wines cited on pp. 184–87, 'The Ultimate Wine Selection', with appropriate dishes. The wines are discussed in the same order as they are mentioned on those pages.

Top St-Émilion Red Wines

These are a glorious match for the Christmas goose, if your taste runs to British traditional; or serve with a rich beef casserole and *pommes mousseline* (potatoes mashed with butter, double cream and milk).

....

Fine Dry White Bordeaux

Pair a Sauvignon Blanc/Sémillon blend with fresh cracked crab, simple lobster dishes or full-flavoured goat's cheeses.

....

Top Sauternes

This is incomparable when sipped very cool as an aperitif, with a handful of almonds; or serve with foie gras or Roquefort cheese.

....

Vintage Champagne

Toast the health of your nearest and dearest before a festive meal; or serve with white meats, such as veal, pork or chicken, in a creamy sauce.

....

Vintage Port

Sip slowly with some freshly cracked walnuts after a meal. If at all possible, ensure that your bottle is decanted and left to air for an hour or so before serving.

....

Late Bottled Vintage Port (LBV)

This is great with Stilton cheese and unsalted crackers.

....

Fine Red Burgundy

This is a terrific choice with an organic roast turkey and all the trimmings on Christmas Day; or serve with classic *coq au vin* casserole.

Great White Burgundy

This is the ideal match with grilled fish, such as sole, turbot or brill. Keep your flavours simple – just a little butter sauce – to complement the roundness of the Chardonnay.

. . . .

Grand Cru Chablis

Here is the perfect match for fresh oysters, or for a selection of seafood served over crushed ice. It is also marvellous with sashimi and sushi.

. . . .

Mature Beaujolais

This delectable expression of the Gamay grape is terrific with rich charcuterie, such as sausage, salami and cold roast pork (or cold turkey).

. . . .

Pouilly-Fuissé

This is a great ice-breaker to serve when guests arrive at a dinner party; or try it with halibut, sea bass or bream dishes. This wine also complements light goat's cheese.

. . . .

Vintage Pink Champagne

This is the ultimate seductive drink for romantic moments; or try it with poached salmon, sea trout or *quenelles de brochet* (pike). It is also perfect with strawberries and cream.

. . . .

Fine Sancerre

Crisp and flinty, this is a good companion to seafood salad or smoked fish, such as salmon, trout or eel. It is also great with fresh, creamy goat's cheese.

. . . .

Vouvray

The creamy notes of this fine Chenin Blanc are a perfect accompaniment to fresh river or lake fish, such as trout and pike; creamy chicken dishes also match it well.

Hermitage

The great Syrah wines of the northern Rhône are full of cracked black pepper flavour, and accompany steak to perfection; try them also with such game as partridge.

. . . .

Condrieu

The Viognier grape has a deep, complex flavour that shows to advantage with grilled fish, such as sea bass, sole and turbot. Condrieu also makes a glorious aperitif for a special occasion, served with lightly salted almonds.

. . . .

Châteauneuf-du-Pape

Deep, dark and intense, this blend of up to thirteen grape varieties deserves a complex dish, such as *bœuf en daube* (beef stew); it is also ideal with any dish that has a rich tomato-and-garlic sauce.

. . . .

Provençal Rosé

This is a good match for fish soups and stews; and for spicy sausage, such as chorizo or merguez.

. . . .

Bandol

The Mourvèdre grape ages well and has a pungency that is good with grilled meats of all kinds, as well as with the regional Provençal speciality, ratatouille.

. . . .

Top Alsace White Wines

Their rich, fragrant style is just perfect with the local dish of choucroute, made with cabbage and various meats; they are also superb with spicy Asian cuisine.

. . . .

Fine German Kabinett Riesling

Try this with trout cooked in butter, with mushroom sauce and dill; or serve with creamy curried chicken and rice.

Beerenauslese Riesling

This is a dessert wine to be savoured slowly at the conclusion of a meal; it is lovely with ripe apricots and with simple crème brûlée.

. . . .

Fine Mosel Riesling

Subtle and light, this wine needs simple food, such as grilled river or lake fish, a creamy mushroom soup, or delicate Thai noodle dishes.

. . . .

Top Barolo

A monster of a wine, this deserves airing for several hours before serving with such game as wild boar, hare or venison. Or try it with salty, hard cheese.

. . . .

Fine Chianti Classico

This deserves a *bistecca alla fiorentina*, which is a good steak; or a game casserole served with creamy polenta.

. . . .

Brunello di Montalcino

A superb match for game, such as pigeon, rabbit, pheasant or quail. It also complements the rich flavour of truffle.

. . . .

Pinot Grigio

This versatile, delicate white will happily accompany many dishes, from poached salmon to Asian cuisine (provided that coriander does not feature in the ingredients, as this can 'kill' a light wine).

. . . .

Fine Vin Santo

Sip a glass of this dessert wine after the meal, with *cantucci* biscuits.

Moscato di Pantelleria

This dessert wine offers another wonderful way to relax after a good repast, with a small selection of candied fruit and sugared almonds to nibble.

....

Top Rioja

Ideal with roast suckling pig, kid or a leg of crispy pork; try it with lamb curry, too. It also teams well with Manchego cheese or Cheddar.

....

Priorat

This is an intense taste experience, which deserves simple grilled meat to accompany it; or a roast wild duck.

....

Fine Ribera del Duero

This goes very well with game birds of all kinds, and venison.

....

Top Cava

This is a great party wine, when served with a selection of tapas.

....

Fine Tokay

Sip this fine dessert wine after the meal, with freshly cracked hazelnuts.

....

New York Cabernet

This is excellent with air-dried ham, or bresaola beef.

....

Top Oregon Pinot Noir

This is just right with roast duck or pheasant.

Fine California Chardonnay

This wine is great with creamy chicken dishes and risotto.

....

Top California Cabernet Sauvignon

Try this with barbecued steak and home-prepared French fries.

....

Fine Zinfandel

This wine is excellent with spare ribs, Mexican dishes and classic meatloaf.

....

Fumé Blanc

This makes a great aperitif – serve it with smoked-fish canapés.

....

Fine California Sparkling Wine

Serve this just as you would Champagne.

....

California Port

Serve after the meal, with some pecan nuts.

....

Top New Zealand Sauvignon Blanc

This is a wonderful wine with all mussel dishes, and asparagus.

....

New Zealand Pinot Noir

The best choice to accompany this is lamb – grilled, roast or even curried.

....

Top Australian Shiraz

The full flavours in this wine are a match for barbecued beef, offal or Mexican dishes.

Fine Australian Cabernet Sauvignon

This is another excellent match for lamb, including kebabs and spicy dishes; or try it with a selection of ripe hard cheeses.

• •

Australian Riesling

This makes the ideal match for 'fusion food' that features chilli, lemongrass or other pungent Asian seasonings.

• •

Top South African Pinotage

This is a marvellous match for any barbecued meat; or try it with game.

• •

Top South African Chenin Blanc

Try this with smoked fish, seafood salads or fish soups.

• •

Fine Chilean Cabernet Sauvignon

This is a great match for any cold-meat selection.

• •

Chilean Sauvignon Blanc

This wine tastes just right with Caesar salad or any creamy fish dish.

• •

Top Argentine Malbec

This is a classic wine to serve with steak and salad.

• •

American or Canadian Icewine

Serve this dessert wine chilled, just by itself — and marvel!

As the sign implies, at Silver Oak Cellars in California they are entirely dedicated to the Cabernet Sauvignon grape in two separate areas of vineyard: one at Alexander Valley and the other in the Napa Valley.

Visiting the Vineyards of Australia and New Zealand

The highlighted areas in the map above show the Australian and New Zealand wine-production regions discussed in this feature.

Australia

At the apex of winter in Europe and the northern hemisphere, there is glorious sunshine in Australia and New Zealand. It is very tempting to make contact with that second cousin and head there to enjoy the beaches, the lively culture and some excellent wine and food. From a slow start in the mid-twentieth century, the wines of Australia have now gained in reputation and popularity, to the extent that more Australian wine than French is now consumed by both Britons and Americans. There always were some excellent bottles in among the quantities of sparkling, dessert and table wine that once represented most of Australia's output. Such wineries as Penfolds (see Interview, pp. 204–205) pioneered the use of such fine varietals as Cabernet Sauvignon as long ago as the nineteenth century; today, even critic Robert Parker is starting to sit up and take notice of quality Australian wine, while collectors are snapping up rare bottles, such as older vintages of Penfolds Grange Hermitage.

So there is wine in quantity and wine of quality to sample, and what is especially appealing about Australian wineries is their enthusiasm for wine tourism. There is none of the European reluctance to spend time with mere *amateurs de vin* here; 'the more the merrier' seems to be the motto. Start your wine odyssey in Sydney, which is just a couple of hours' drive from the Hunter Valley vineyards, where the Australian wine story began. This is a region rich in history as well as having dozens of great wineries and restaurants; and it is even possible to take your wedding vows in the Hunter Valley Gardens, or play golf amid the vines belonging to top player Greg Norman.

One of the most significant wine players in the Hunter Valley is Tyrrell's, established in 1858. Its winemakers have battled the extreme local climate, with its hailstorms, searing heat and heavy humidity, to produce some fine red and white wines, notably Chardonnay (Tyrrell's was among the earliest to plant this variety) and Semillon, once used to make sweet dessert wines but now

Top: Vines stretch into the distance in the Yarra Valley, in the Victoria wine region.

Above: A roadside sign at Coonawarra, famed for its fine Cabernet and Merlot wines, displays typically Aussie wit.

ABOVE: Coonawarra is an epicentre for the Australian wine industry and home to such household names as Jacob's Creek.

OPPOSITE: The Hunter Valley of New South Wales is where it all began in Australia, with such now-familiar names as Tyrrell's and Rosemount.

ABOVE: Australia's kangaroos are renowned for their fearless approach to man; be it golf course or vineyard, they are seen out and about.

LEFT: Considerable skill and hard work are required to harvest grapes by hand, and the intense heat of Australia makes it an extra effort.

a truly distinctive Australian dry white. There is also some classic, spicy Shiraz that is full of character.

Move on to Melbourne, a magnet for gourmets and wine enthusiasts, with a great café culture. The state of Victoria is strongly associated with the Brown Brothers wines, and their Milawa vineyard is well worth visiting. There is truly something for everyone here: wines range from low-alcohol, easy drinking to fine, rich dessert wine. You could also drive out to the lovely Yarra Valley, a restful place not far from the city, where you can sample the fine sparkling wines at Domaine Chandon, an offshoot of the Moët & Chandon Champagne house; it has a beautifully sited tasting room overlooking the vines. At De Bortoli, where there is an emphasis on wine and food pairing and a Yarra Valley Wine School, you can enjoy further wine tasting.

The equivalent to California's Central Valley (the 'workhorse' region of California, producing huge quantities of grapes for wine and for the table) should be South Australia, where endless litres of straightforward drinking wine are blended annually and sold worldwide. Yet the comparison is not precise: there are also very fine wines produced alongside, at Penfolds, Jacob's Creek and Peter Lehmann. For an interesting tasting, visit Shaw + Smith's winery, where you can sample their Shiraz, Sauvignon Blanc and Chardonnay with some local cheese. Over in the McLaren Vale, there are some fascinating, experimental vineyards, as well as such established stars as d'Arenberg, a family business with a wide selection of wines, many made with Rhône varietals, and including some fine 'stickies' as a homage to the early days of winemaking in Australia, when dessert wines were hugely popular.

In Western Australia, there is much excitement among wine connoisseurs over the wines made in the Margaret River region. Two pioneering wineries that are worth a visit are Cullen Wines (with grapes grown biodynamically) and Leeuwin Estate, famed for its complex, sophisticated Chardonnay.

New Zealand

This small country is a must-see for anyone who appreciates fine wine: apart from its stunning physical beauty, there is no doubt that the Sauvignon Blanc and Pinot Noir made here have a very specific charm and style that seem inimitable. There are now some 550 wineries, many making tiny quantities of great wine, and a tour is the perfect way to appreciate this diversity and quality.

Starting in Auckland, where most international flights land, take a look around nearby vineyards, such as the excellent Villa Maria Estate, the second-largest wine company in the country and a source of truly reliable, well-made wines, most sealed with a screwcap. This replacement for the traditional cork was pioneered here, and taken up by many other New Zealand producers faced with that long export journey to market and the risk of wine spoilage on arrival. The experiment has been a resounding success, and winemakers all over the world are now following suit.

Also on North Island, head out east to Gisborne and visit the biodynamic Millton Vineyard, producing fine Riesling, Chardonnay and Chenin Blanc. As a contrast, spend some time at the Lindauer/Montana winery, where you can sample some of the most successful New Zealand branded wine from this leading company, as well as very appealing Lindauer fizz. There are also interesting vineyards in the Hawke's Bay region, which is home to many plantings of Cabernet Sauvignon.

The glorious South Island is the place to go for iconic Sauvignon Blanc, made in the Marlborough region; perhaps the most famous name is Cloudy Bay, but there are also great examples from Villa Maria and Montana, not to mention a host of small, individualistic producers who do not export at all. For Pinot Noir, Central Otago is a must (see note on Sam Neill in 'July', p. 104); the area produces wine of this variety that has gained an international reputation. It is also a region famed for its rugged scenery, so if all that serious sipping and swirling at such wineries as Gibbston Valley or the aptly named Mt Difficulty leave you feeling fidgety, burn up some of your excess energy in hiking, rafting, kayaking or even bungee jumping. Who said wine tourists had to be sedate?

BELOW: Marlborough's first commercial vineyard was planted in 1973; today the region yields half of New Zealand's production.

BOTTOM: Gisborne, a warm and humid region in New Zealand's North Island, is home to some fine Chardonnays.

Peter Gago

Chief Winemaker, Penfolds, Australia

Penfolds was established in 1844 by Dr Christopher Rawson Penfold (see p. 74). The company has gone on to become one of Australia's most distinguished winemakers.

penfolds.com

Could you describe your personal style of winemaking – and give us a little background on your own history?
I'm very fortunate as Penfolds Chief Winemaker to have the luxury of working with a dedicated team with many decades of winemaking expertise. Steve Lienert, our Senior Red Winemaker, has more than thirty years' experience crafting Penfolds wines. His predecessor, John Bird, still consults for us today; he celebrated fifty consecutive vintages with Penfolds in 2009. Winemakers Andrew Baldwin and Kym Schroeter both have well over twenty years' experience. Winemaking at Penfolds is about a collaborative team effort, and while I have my hand on the tiller, I couldn't achieve what I have without such a wealth of collective knowledge to guide me along. Ultimately, the philosophy of Penfolds is built on quality, longevity and commitment to style. We like to regard Penfolds as the world's largest boutique winery!

Do you believe Australian wines are good investments?
The secondary market [auction or private sale] is very much alive and well in Australia. At a recent auction, an Imperial (6 litres) of 2004 Grange fetched AU$30,000. The 1976 Koonunga Hill Shiraz Cabernet amply demonstrates that cellaring wine is not necessarily the pursuit of the wealthy. Originally released at $2 a bottle, it has developed into a magnificent mature wine and can fetch more than $300 at auction today.

Penfolds offers a unique after-sales service, the Red Wine Recorking Clinics. Since the inception of the programme in 1991, Penfolds has recorked more than 95,000 bottles of wine across the globe. This free service offers consumers the opportunity to have their wines checked and assessed by a Penfolds winemaker. Good bottles are certified, recapsuled and logged in our database. Valuations can be obtained from an independent wine investment expert at every clinic.

How do you see the future for the company, and the wines of Australia?
The Australian wine industry is facing a range of challenges. The influence of climate change is already self-evident. The past ten years of drought have encouraged viticulturalists to explore more efficient practices and to experiment with 'alternative' grape varietals while seeking to improve the overall standard of the existing Australian wines on offer.

Could you nominate a personal favourite from the wines you produce?
I've always been especially fond of Penfolds Bin 389, a wine originally created in 1960 by Max Schubert. It is a blend of Cabernet Sauvignon and Shiraz sourced from some of our finest vineyards. Sometimes referred to as 'Baby Grange' (but at a fraction of the price of its better-known stablemate), Bin 389 highlights the benefit of multi-regional blending across the two varietals. The glorious structure of Cabernet Sauvignon is bolstered by the luscious intensity of Shiraz to create a wine of distinction. Good vintages of Bin 389 have the capacity to age gracefully for upwards of twenty-five years, but also offer immediate accessibility in youth.

Finally, could you give the readers a couple of reasons why it's worth coming to Australia for a vineyard visit?

Penfolds Yattarna derives its name from an Aboriginal word meaning 'little by little', reflecting the attention to detail needed to create this sophisticated white wine, made with 100 per cent Chardonnay.

Australia is a vast country blessed with an abundance of diverse, beautiful wine regions offering something of interest to even the most discerning gourmet traveller. A thriving tourism industry has grown symbiotically alongside the vineyards and wineries across all of Australia's premier regions. Of particular delight are the numerous restaurants showcasing quality local produce matched to regional wines. In recent years, a number of boutique luxury hotels have opened alongside the traditional 'Bed & Breakfast' cottages that have long been a favourite among locals. Most Australian wineries offer cellar-door tastings, more often than not for free. At our winery at Nuriootpa in the Barossa Valley, visitors can partake in a 'Make your own blend' exercise in our winemakers' tasting room. This unique experience allows you to test your blending skills by creating your own version of the Bin 138 Barossa Valley Grenache Shiraz Mourvèdre. Or visit our Magill Estate Restaurant, a fifteen-minute drive from Adelaide's Central Business District, nestled in a historic vineyard alongside a working winery – Penfolds' spiritual home.

December is a glorious, sunny month in the Western Cape, so South Africans flock to the annual Franschhoek Cap Classique and Champagne Festival early in the month, where a range of fine table and sparkling wines are on show with gourmet food: franschhoek.org.za

In Hobart, Tasmania, the Taste Festival runs from late December to early January, offering a chance to sample the best of Australia's food and drink: tastefestival.com.au

The rest of the wine world rests a little now, making the most of the Christmas season and the opportunity to serve some excellent vintages with the seasonal feast and at New Year parties on 31 December. With a glass of Champagne (or its regional equivalent) in hand, we can truly welcome in the New Year.

A barrel shows a cellarmaster's log of the condition of the wine it contains; this record will also show the date of racking (pumping the wine from one container to another, to separate it from its lees). Every cellarmaster has his or her particular notation system.

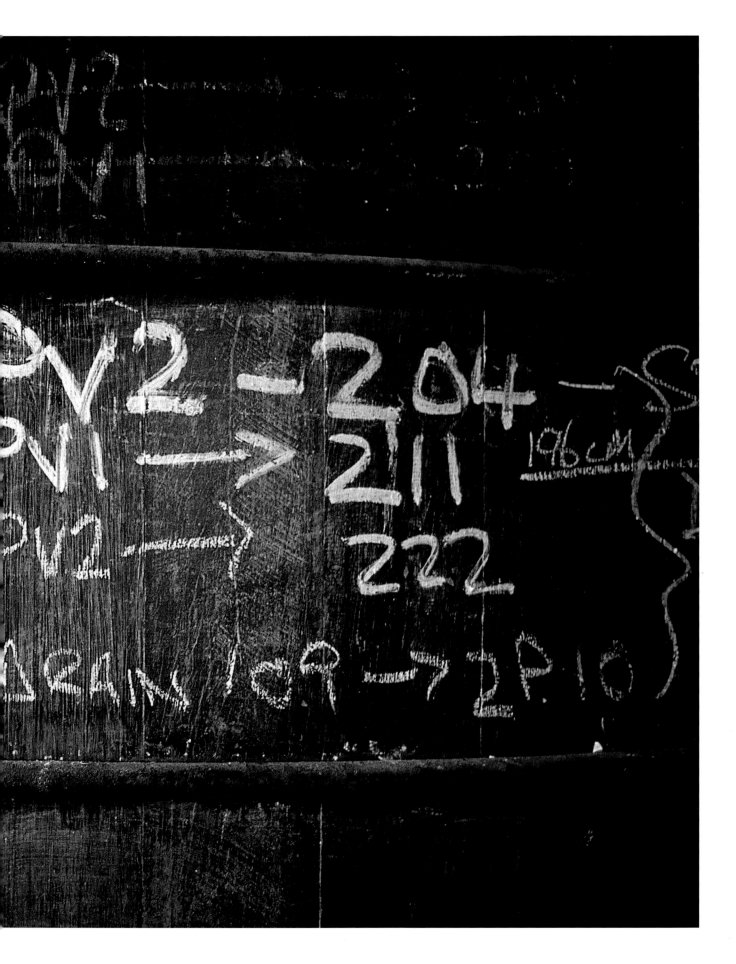

The Language of Wine

···

acidity

A vital element in any good wine, naturally present, although it may be augmented in very hot climates with such additives as citric acid.

adega

The Portuguese word for 'winery'.

ageing

Also called 'maturation', an important process for fine wines, especially reds.

agriturismo

Holiday accommodation in Italy in a farmhouse location. Such places usually offer the chance to sample food and wine produced on the farm.

alcohol

This is expressed as a percentage of the wine volume, typically around 11%–15%.

alcoholic fermentation

A natural process whereby the sugar in the grapes, mixed with natural yeasts found on the skins, converts to alcohol (ethanol) and gives off carbon dioxide (CO_2). Modern winemakers tend to use prepared yeasts and control the temperature to ensure a reliable product.

American Viticultural Area (AVA)

The defined 'appellation' zones in US wine-producing regions.

Appellation d'Origine Contrôlée (AC or AOC)

The top category for French wines, governed by regulations that apply to such factors as yield, grape variety, location of vineyards and alcohol level.

assemblage

See **blending**.

Auslese

A category of German wine, in which the grapes are riper and the wine a little sweeter than a typical **Kabinett**.

azienda agricola

An Italian wine estate or farm; usually means the grapes are grown on the property.

barrel fermented

A term used to describe wine that has been fermented in oak barrels to give additional depth of flavour.

barrique

A classic French wine oak vat: the *barrique bordelaise* holds 225 litres (49.5 gallons).

Beerenauslese

A higher grade of sweetness for a German wine than **Auslese**; grapes that create Beerenauslese wines are affected by ***Botrytis cinerea***, known as *Edelfäule* in German and 'noble rot' in English.

biodynamic viticulture

The process of growing grapes according to the phases of the moon and the movement of the planets, following principles established by Rudolf Steiner (see p. 62); increasingly popular, even for fine-wine production.

Blanc de Blancs

French for 'white from whites'; still or sparkling white wine made with white grapes only (the juice of red grapes is generally also white, so Pinot Noir red grapes may be used in making Champagne, for example).

Blanc de Noirs

French for 'white from blacks'; still or sparkling white wine made with the juice of red grapes only; there is no skin contact. Most commonly seen on Champagne labels, as it has a legal significance in this region.

Blanquette de Limoux

A sparkling white wine from the Limoux region of France. The wine can include Mauzac (which must be the principal grape in its make-up), Chardonnay and Chenin Blanc. It has a unique, apple-peel flavour and grassy aroma.

blending

Known as *assemblage* in French; an essential process for many traditional wines, to create either a familiar 'house' style or to improve wine in a poor vintage.

bodega

The Spanish word for 'winery'.

Botrytis cinerea

The Latin name for 'noble rot', a mould that affects grapes used in Sauternes and other dessert wines, to give a distinctive earthy, rich style and dry finish.

bouquet

A term used to describe the deep, intricate scents that evolve in a wine as it ages. These usually develop after the wine is bottled, and are linked to the way in which the wine has been fermented and processed. They are different from the scents that arise naturally from the fruit in the wine, which are known as aromas.

brut

A dry sparkling wine, especially Champagne, with only a very small **dosage** of sugars.

carbonic maceration

A winemaking method used to extract maximum fruit and little **tannin**, for example in Beaujolais. Whole bunches go straight into a closed fermenting vat and crush

naturally, yielding intense **varietal** flavour.

cane

A common term for the bare stems of the vine, before or after pruning.

cave

A French wine cellar.

chai

The French term for an above-ground storage building for wine.

Champagne Method

See **Méthode Champenoise**.

chaptalization

The process of adding sugar during fermentation to increase the alcoholic strength of a wine; necessary and legal in some marginal climate zones where grapes struggle to ripen.

Charmat Method

Also called 'tank method', or *cuve close* in French, this is a way of producing sparkling wine by carrying out the secondary fermentation process in a large vessel or 'tank' instead of in the individual bottle (as in the **Méthode Champenoise**).

château (Ch.)

The French word for 'castle'; a Bordeaux name for a wine estate, also used in other French regions.

claret

The traditional name for red Bordeaux, referring to *clairet*,

an old-fashioned, pale-red wine still produced on a small scale in the region.

clone

A grape variety selected and bred for a particular quality; for example, a specific Pinot Noir that ripens more readily or gives more colour than others.

clos

A walled vineyard (as in Clos de Vougeot in Burgundy) that shelters grapes to result in improved quality; may also be used without this literal meaning.

corked wine

Wine that is spoilt, not directly by a rotting cork, but because of the presence of 2,4,6-trichloroanisole (TCA) in a wine; this may be avoided by the use of a screwcap.

cosecha

The Spanish word for '**vintage**'.

Côte

The French word for 'hill' or 'slope'; better vines tend to grow on higher land.

crémant

A creamy style of sparkling wine with a little less pressure in the bottle than Champagne; made in Alsace, the Loire Valley, the Jura and other French regions.

crème de cassis

A liqueur made from blackcurrants; can be mixed with white wines or Champagne to make Kir.

crème de myrtilles

A liqueur made from bilberries; can be mixed with white wines or Champagne.

crianza

A Spanish word referring to the youngest oak-aged wine that may be sold: two years old for a red wine, one year old for a white.

cru

The French word for 'growth'; in wine parlance, a particular vineyard site or estate known for its quality.

Cru Bourgeois

Better-value wines from good Bordeaux estates, just below the level of the top **Crus Classés**; *see also* **Grand Cru**.

Crus Classés

Wines from Bordeaux that have been categorized according to five quality levels, or growths, with **Premier Cru** ('first growth') being the highest; all are fine wines, but some classifications are out of date (dating to 1855 in the Médoc). For example, Ch. Lynch-Bages is rated a fifth growth, yet is generally reckoned to be of higher quality than many third growths. *See also* **Grand Cru**.

cuve close

See **Charmat Method**.

cuvée

A single vat of wine; several vats go to make a blend, which, confusingly, may also be called a cuvée.

disgorging

The process of expelling the sediment from a bottle of sparkling wine made by the **Méthode Champenoise**; the process of **remuage** is used to shake down the sediment gently for a period of time beforehand.

demi-sec

A French term meaning medium dry, but confusingly applied to wines that are medium sweet.

Denominación de Origen (DO)

The Spanish version of the French **Appellation d'Origine Contrôlée** rules.

Denominazione di Origine Controllata (DOC)

The Italian version of the French **Appellation d'Origine Contrôlée** rules; there is also a higher category called Denominazione di Origine Controllata Garantita (DOCG), for certain top wines.

dosage

The amount of natural sweetening that is added to a Champagne or sparkling wine before it is bottled.

doux

The term used to describe a sweet style of still or sparkling wine.

Eiswein

A German or Austrian sweet white wine made from grapes that have frozen on the vine.

filtering

The process of removing any residue from fermentation (such as yeast cells) before bottling.

fine mousse

A French term describing the steady stream of small bubbles produced by the **Méthode Champenoise**, as opposed to the **Charmat Method**.

fining

The process of causing protein deposits to separate from wine to ensure clarity, by using – among other substances – egg white or bentonite.

finish

A wine-tasting term used to summarize the complex impressions left on the palate after tasting a wine.

fino

The driest style of sherry from Spain, served chilled.

first growth

An often-used term for top Bordeaux red wines; *see* **Crus Classés**.

flute

The preferred style of glass for tasting or drinking Champagne to showcase the *fine mousse*.

flying winemaker

The practice, begun in the 1970s, of trained Australian or American winemakers 'parachuting' into a winery at vintage time to assist with making the wine; many work in regions as varied as Portugal, Moldova and Chile.

fortified wine

A style of wine in which the base wine has been augmented with grape spirit, stopping fermentation and leaving residual sweetness; examples include port, sherry and marsala.

frizzante

The Italian term for sparkling wine, less fizzy than **spumante**.

gobelet

A system of vine training whereby vines are trimmed so that they grow in a shape similar to a small bush or shrub. This is traditional in many areas, such as Beaujolais, and is especially suitable for older vines or those grown in windy vineyards.

Grand Cru

A term used primarily in Bordeaux, Burgundy, Chablis and Alsace; indicates the higher quality vineyards in these regions, but is not a guarantee of finesse. The producer's or shipper's name should also be considered when choosing a wine.

grappa

An Italian clear spirit, made with the **pomace** that remains after fermentation. This fiery drink varies in quality; Nonino is a top producer.

growth

A term often used by wine lovers when referring to a **cru** from Bordeaux; hence the phrase 'classed growth wine'; *see also* **Crus Classés**.

Halbtrocken

A German wine style tending towards dry, compared with a typical **Kabinett**.

Icewine

The American and Canadian version of **Eiswein**.

Indicazione Geografica Tipica (IGT)

The Italian version of French **Vin de Pays**; wines that are good value and typical of a region.

Kabinett

The simplest version of a quality **Prädikat (QmP)** wine in Germany.

Late Bottled Vintage (LBV)

A ruby port that is kept in barrel for between four and six years; it is usually lighter bodied than vintage port.

lees

What remains in the vat or barrel after fermentation. Some wines, such as Muscadet and Champagne, are left on their lees to add extra flavour.

lutte raisonnée

The French term for 'reasoned struggle'; describes a method of growing vines whereby the use of chemicals is kept to a minimum, and natural fertilizers

are used, so that the vines find their own strength.

malolactic fermentation

A secondary process in winemaking whereby the tart 'appley' malic acid naturally present in wine breaks down to softer lactic acid. It will often occur naturally, but modern technologies are employed to control the process. The result is a more palatable, attractive wine.

maturation

See **ageing**.

Meritage

An American term for a Bordeaux-style blended wine, made with traditional varieties.

Méthode Champenoise

The method for making sparkling wine by secondary fermentation in the actual bottle you are purchasing. Recent regulations require that no other region than Champagne may use this phrase.

Méthode Traditionelle

The method for making sparkling wine that is derived from practices first used in Champagne. Only wine producers in that region of France may use the phrase **Méthode Champenoise**.

moelleux

The French word for 'mellow', a sweeter style of wine.

mousseux

The French word for sparkling wine.

must

Crushed grapes and juice prior to fermentation.

négociant

A French wine merchant who buys, sells and may also blend wines.

noble rot

See **Botrytis cinerea**.

oenology

The study of wines and winemaking.

parcelle

A plot of vines in France (French inheritance laws require that vineyards are subdivided among many heirs).

passito

An Italian term for wine made with semi-dried grapes, such as **Vin Santo**.

pétillant

A French term for a lightly sparkling wine.

phylloxera

A vine pest that gnaws the roots of **Vitis vinifera**, resulting in these vines having to be grafted on to American rootstock. The American vines used as rootstock are resistant to this pest as they are of the species *Vitis labrusca*.

pomace

Wrung-out grape skins; the residue after the grapes have been pressed.

Prädikat

A German wine distinction that is awarded on the basis of increasing grape ripeness; categories include (in ascending order of sweetness) **Kabinett**, **Spätlese**, **Auslese**, **Beerenauslese**, **Eiswein** and **Trockenbeerenauslese**.

Premier Cru

A French term that indicates top-quality wine; may be used for fine wines in various regions of France, including Bordeaux, Chablis and Burgundy.

primeur

The first wines to be offered to market; buying *en primeur* is popular among wine collectors.

punching down

A process in the making of red wine in which the cap of grape skins that forms on top of the fermenting wine is plunged down into the **must** so that it does not dry out and stays in contact with the evolving wine.

QbA (Qualitätswein bestimmter Anbaugebiete)

The German category for quality wines from designated regions.

QmP (Qualitätswein mit Prädikat)

A higher quality category for German wines.

quinta

The Portuguese word for 'farm' or 'estate'.

racking

The process of transferring wine from one vat to another, to clear and settle after fermentation.

récoltants-manipulants

Growers of Champagne grapes who produce the Champagne entirely from their own vineyards.

remuage

The French word referring to the gradual movement of sediment in a Champagne bottle down to its neck, by gently shaking the bottle in a special rack, either by hand or by machine; the wine is then **disgorged** before corking.

reserva

Spanish wines that have been selected and aged for specific periods. The Italian equivalents are called *riserva*.

Sekt

The German term for sparkling wine; the best are made by the **Méthode Traditionelle**.

Spätlese

An off-dry German wine made with riper grapes.

spumante

The Italian term for sparkling wine.

sur lie

A wine that has been allowed to rest on its **lees** after fermentation, to gain extra flavour.

Syrah/Shiraz

Syrah is a classic, 'noble' grape of southern France that is now widely planted in other wine regions. The Shiraz grape, a close relative of Syrah, is grown in Australia and has been a huge commercial success. As a result, some winemakers from other regions have 'borrowed' the name Shiraz (perfectly legally) to give their wines added saleability.

Tafelwein

German table wine, the lowest quality level.

tannins

Naturally occurring harsh, tough elements in wine, especially reds, derived from the skins, pips and stems of grapes; vital for the **ageing** process, when they soften and alter in flavour.

terroir

The French term for a vineyard area that is best suited to certain grapes, owing to its climate and soil type.

trocken

The German word for 'dry'.

Trockenbeerenauslese

Confusingly, not a dry wine but an intensely sweet German or Austrian dessert wine made with *Botrytis* grapes; it has a clean, crisp finish.

varietal

A wine named for the grape or grapes that dominate the blend, but it may consist of 100 per cent of a single variety.

vieilles vignes

The French term for 'old vines' (at least thirty years old); these often produce high-quality wines, which may be higher in alcohol.

vigneron

The French term for a winemaker who cultivates grapes.

Vin de Pays

The French wine category for typical, regional wines.

Vin de Plaisir

The French term for a fine wine suitable for extended appreciation in a meditative moment.

Vin de Table

French table wine; the lowest quality level.

Vin Doux Naturel

A French fortified wine served as an aperitif or with dessert.

Vin Santo

A classic Italian fine dessert wine from Tuscany and Umbria, made with grapes dried on racks or mats; the ensuing wine is aged for up to ten years in barrel. The term is sometimes 'borrowed' for inferior wines with added sweetening, so study the label for origin.

vinification

The process of transforming grapes into wine.

Vino da Tavola

Italian table wine. Experimental winemakers also use this category to indicate fine wines made with non-classic grape varieties, such as Merlot in Tuscany.

vintage

The wine harvest; also wine from a single year. Champagne and port do not offer vintage wines in every year; this is dependent on quality.

viticulture

The process of cultivating grapes for wine.

Vitis vinifera

The species of vine that includes all the classic wine varieties. Other species include *Vitis labrusca* and *Vitis rupestris*, both native to North America. Examples of *vinifera* include Cabernet, Chardonnay and Pinot Noir, all European native varieties brought from Italy to other regions by the Romans.

yield

The amount of wine obtained from a certain quantity of grapes; in regulated areas, fine wines' yield is strictly controlled, and pruning is vital.

Zin

An affectionate nickname for the Zinfandel grape of California, a cousin to the Italian Primitivo.

WEBSITES FOR WINE LOVERS

General Information

bbr.com
   British fine wine merchants Berry
   Bros. & Rudd.

decanter.com
   Informative website from the top UK
   wine magazine.

erobertparker.com
   Internationally recognized US wine-
   taster's site: accessible on subscription.

jancisrobinson.com
   Top UK wine-writer's site: accessible
   on subscription.

laithwaites.co.uk
   The flagship site for the acknowledged
   master of mail-order wine.

wineanorak.com
   Useful UK-based site from wine
   journalist and author Jamie Goode.

wine-pages.com
   Long-established online site from
   wine writer Tom Cannavan.

wine-searcher.com
   The place to locate any wine from
   around the world.

winespectator.com
   Very helpful site associated with the
   major US magazine.

Wine Education

brucecasswinelab.com
   Excellent wine school in the San
   Francisco Bay Area.

wineeducation.org
   Site devised by wine educator
   Stephen Reiss, with a less formal
   approach to wine education.

wineeducators.com
   Home site of the Association
   of Wine Educators (AWE), a UK
   group of independent, professional
   wine educators.

wset.co.uk
   Site of the Wine & Spirit Education
   Trust (WSET), the official training
   ground for UK wine professionals.

Regional Sites

ARGENTINA
   winesofargentina.org

AUSTRALIA
   wineaustralia.com

AUSTRIA
   austrian.wine.co.at

BULGARIA
   bulgarianwines.com

CANADA
   winesofcanada.com

CHILE
   winesofchile.org

CHINA
   wines-info.com/en/

FRANCE
   frenchwinesfood.com
   terroir-france.com
   Alsace   vinsalsace.com
   Bordeaux   bordeaux.com
   Burgundy   bivb.com
   Champagne   champagne.fr
   Languedoc   languedoc-
                     wines.com
   Loire   loirevalleywine.com
   Provence   vinsdeprovence.com
   Rhône   vins-rhone.com

GERMANY
   germanwine.de
   winepage.de

GREECE
   allaboutgreekwine.com

HUNGARY
   winesofhungary.com

ISRAEL
   israelwinecompany.com

ITALY
   italianmade.com/wines/home.cfm
   Tuscany   chianticlassico.com

LEBANON
   chateaukefraya.com
   chateaumusar.com.lb

NEW ZEALAND
   nzwine.com
   tizwine.com

PORTUGAL
portugal-info.net/wines/general.htm
Madeira   madeirawineguide.com
Port   portwine.com

SOUTH AFRICA
wosa.co.za

SPAIN
winesfromspain.com
winesfromspainusa.com
Ribera del Duero riberadelduero.es
Rioja   riojawine.com
Sherry   sherry.org

SWITZERLAND
wine.ch

UNITED KINGDOM
englishwineproducers.com

UNITED STATES
allamericanwineries.com
California   napavintners.com
                wineinstitute.org
New York   newyorkwines.org
Oregon   oregonwine.org
Washington   washingtonwine.org
Virginia   virginiawine.org

URUGUAY
travelenvoy.com/wine/uruguay.htm

Wine Tours
· ·

abcwine.it
Tuscan wine tours.

amazingmendoza.com
Visits to Argentine vineyards.

cellartours.com
Specialists in tours of wineries in Spain and Portugal.

gourmetontour.com
Eating and drinking in Europe, the USA, Asia and Morocco.

intouchtravel.com
Personalized approach to touring.

luxurywinetourism.com
Bernard Magrez's Bordeaux agency, offering a range of wine-tourism activities.

tastesa.com.au
Wine tours in South Australia.

uniworld.com
River cruises on the Rhine and Danube, including wine tastings.

wine-tours-france.com
Tailor-made French wine tours.

winetours.co.uk
Well-established wine-tour company Arblaster & Clarke.

winetours.co.za
Visiting the Cape vineyards.

Fairs and Festivals
· ·

AUSTRALIA
crushfestival.com.au
melbournefoodandwine.com.au
southcoastwineshow.blogspot.com

CANADA
niagarawinefestival.com

ENGLAND
glynde.co.uk
londonwinefair.com

FRANCE
concours-des-vins.com
fetedesvins-anjou.fr
hospices-de-beaune.com
vinexpo.com

ITALY
fieradeltartufo.com
meranowinefestival.com
sommelier.it
vinitaly.com
vitignoitalia.it
wineshow.it

NEW ZEALAND
hawkesbay.com
wine-marlborough-festival.co.nz

SOUTH AFRICA
capewinemakersguild.com
wineroute.co.za

UNITED STATES
liwinterfest.com
oregontrufflefestival.com
vawineshowcase.org
wayoutwineries.org

PREVIOUS PAGE: The vine as art: sculptural posts and wire-frame pruned vines at the Ken Forrester vineyards in South Africa.

OPPOSITE: Taking a wine tour that features tastings is a fascinating combination of education and entertainment, and an excellent way of gaining an in-depth knowledge of a region.

Two views of the vineyards in
Alsace, France. The wines, from
steep vineyards in the lee
of the Vosges mountains,
combine French and German
influences and feature such
grapes as Riesling and Sylvaner.

Acknowledgements

This book is dedicated to Colin Parnell (1934–2010), founding editor of *Decanter* magazine and a friend and mentor early in my career.

I would like to thank the fourteen distinguished winemakers and company owners who kindly agreed to be interviewed for the twelve chapters of *The Wine Year*: Alberic Bichot; Gianluca Bisol; Bob Blue; Ken Forrester; Peter Gago; Elisabetta Geppetti; Dr Clemens Kiefer; Chiara Lungarotti; Joe Macari; Bernard Magrez; Scott Osborn; Antonio Pinilla; Mario Pablo Silva; and Stéphane Tsassis.

I also appreciate the help I received from the following while researching this book: Patricia Parnell; Jill Norman; Sue Glasgow at Spear Communications; Rosamund Hitchcock at R&R Teamwork; Kate Sweet at Limm PR; Emma Wellings PR; Nina Plowman at Wild Card PR; Lucy Richardson at Phipps PR; Richard Bampfield MW; Ursula Thurner of Thurner PR, Florence, Italy; Bibendum Wine Limited; Lisa Duckenfield at Cellar Trends; Bethan Wallace at Clementine Communications; Bryony Wright at Proven PR; Fran Draper at Rumpus Communications; Kate Diggle at Laithwaites; Julia Trustram Eve; Fiona Sanderson at The Luxury Channel; Dacotah Renneau; David Lindsay at Lindsay May PR; Mo Milton at Nelson Bostock Communications; Susan Spence at the New York Wine & Grape Foundation; Judi Betts in Long Island; and my good friend Bruce Cass in California. Not to mention the constant support of my wine-tasting family: Aldwyn, Theo, Zoë and Claudia.

Rosalind Cooper, 2010

Picture Credits

INDEX

Rosalind Cooper is an experienced journalist and writer on wine. She has published several books on the subject, and was formerly Deputy Editor of *Decanter* magazine and Editor of *Wine Times*, the magazine of the Sunday Times Wine Club.

COVER: Harvest-ready Carmenère grapes at Viña Casa Silva, Chile (see p. 71).

BACK COVER INSETS, LEFT TO RIGHT, FROM TOP: Snow-covered Torres vine terraces in Catalonia, Spain (see p. 50); snow and ice on a dormant Bordeaux vine (see p. 13); the growing vine buds; spring flowers in a Bordeaux vineyard; a wasp attracted by the sugar in fully ripe grapes (see p. 134); Ken Forrester's homestead in Stellenbosch, South Africa (see p. 100); grapes being picked in New York State (see p. 159); juice extraction gets under way in Champagne (see p. 22); Rioja vineyards, Spain (see p. 50); the vine in autumn; a celebratory table setting (see p. 108); wine ageing in oak barrels (see p. 151).

PAGE 1 and FRONTISPIECE: Scenes from Bordeaux vineyards (see p. 142).

First published 2010 by

Merrell Publishers Limited
81 Southwark Street
London SE1 0HX

merrellpublishers.com

Text copyright © 2010 Rosalind Cooper
Illustrations copyright © the copyright holders; see p. 219
Design and layout copyright © 2010 Merrell Publishers Limited

British Library Cataloguing-in-Publication Data:
The wine year.
1. Wine and wine making. 2. Wine districts – Guidebooks.
3. Food and wine pairing.
I. Title
641.2'2-dc22

ISBN 978-1-8589-4514-9

Produced by Merrell Publishers Limited
Designed by Alexandre Coco
Project-managed by Lucy Smith and Marion Moisy
Proof-read by Kate Michell
Indexed by Hilary Bird

Printed and bound in Hong Kong